PERFORMING TOURIST PLACES

New Directions in Tourism Analysis

Series Editors: Kevin Meethan, University of Plymouth
Dimitri Ioannides, Southwest Missouri State University

Although tourism is becoming increasingly popular as both a taught subject and an area for empirical investigation, the theoretical underpinnings of many approaches have tended to be eclectic and somewhat underdeveloped. However, recent developments indicate that the field of tourism studies is beginning to develop in a more theoretically informed manner, but this has not yet been matched by current publications.

The aim of this series is to fill this gap with high quality monographs or edited collections that seek to develop tourism analysis at both theoretical and substantive levels using approaches which are broadly derived from allied social science disciplines such as Sociology, Social Anthropology, Human and Social Geography, and Cultural Studies. As tourism studies covers a wide range of activities and sub fields, certain areas such as Hospitality Management and Business, which are already well provided for, would be excluded. The series will therefore fill a gap in the current overall pattern of publication.

Suggested themes to be covered by the series, either singly or in combination, include - consumption; cultural change; development; gender; globalisation; political economy; social theory; sustainability.

Also in the series

The Challenge of Tourism Carrying Capacity Assessment:
Theory and Practice
Edited by Harry Coccossis and Alexandra Mexa
ISBN 0 7546 3569 4

New Directions in Rural Tourism
Edited by Derek Hall, Lesley Roberts and Morag Mitchell
ISBN 0 7546 3633 X

Performing Tourist Places

JØRGEN OLE BÆRENHOLDT
Roskilde University, Denmark

MICHAEL HALDRUP
Roskilde University, Denmark

JONAS LARSEN
Roskilde University, Denmark

JOHN URRY
Lancaster University, UK

ASHGATE

Published by
Ashgate Publishing Limited
Gower House
Croft Road
Aldershot
Hants GU11 3HR
England

Ashgate Publishing Company
Suite 420
101 Cherry Street
Burlington, VT 05401-4405
USA

Ashgate website: http://www.ashgate.com

British Library Cataloguing in Publication Data
Performing tourist places. - (New directions in tourism
   analysis)
   1.Tourism - Social aspects 2.Tourism - Social aspects -
   Denmark 3.Resorts - Denmark
   I.Bærenholdt, Jørgen Ole
   338.4'791

Library of Congress Cataloging-in-Publication Data
Performing tourist places / Jørgen Ole Bærenholdt ... [et al.].
       p. cm. -- (New directions in tourism analysis)
   Includes bibliographical references and index.
   ISBN 0-7546-3838-3
       1. Tourism. 2. Tourism--Social aspects. I. Bærenholdt, Jørgen Ole. II. Series.

   G155.A1P3758 2003
   338.4'791--dc22

2003063728

ISBN 0 7546 3838 3

Printed and bound in Great Britain by MPG Books Ltd, Bodmin, Cornwall

Contents

List of Figures

List of Tables

List of Authors

Jørgen Ole Bærenholdt, Associate Professor, Department of Geography and International Development Studies, Roskilde University, Denmark.

Michael Haldrup, Associate Professor, Department of Geography and International Development Studies, Roskilde University, Denmark.

Jonas Larsen, Research Fellow/Teaching Assistant, Department of Geography and International Development Studies, Roskilde University, Denmark.

John Urry, Professor, Department of Sociology, Lancaster University, UK.

Preface

This book presents the results from the three-year research project 'Tourism Practices and the Production of Destinations – Representations, Networks and Strategies' based within the Department of Geography and International Development Studies, Roskilde University. The project was one of 14 projects carried out at the Tourism Research Centre of Denmark, 1999-2003. The Research Centre was established as a cooperation between Roskilde University (Department of Geography and International Development Studies and Department of Social Sciences and Business Economics), Copenhagen Business School (Department of Management, Politics and Philosophy), and Centre for Regional and Tourism Research (former Research Centre of Bornholm), funded by the Danish Social Science Research Council.

Although the book is a collective work, the spatial division of fieldwork has meant that Jørgen Ole Bærenholdt and Jonas Larsen have concentrated on Bornholm (chapters 3, 5 and 6) while Michael Haldrup undertook fieldwork on Northern Jutland (chapters 4, 6 and 7). Chapter 2 is partly based on findings by Wolfgang Framke, Flemming Sørensen and Per-Åke Nilsson. However, Jørgen Ole Bærenholdt has mainly drafted chapter 2 with substantial contributions from Wolfgang Framke; an effort we would especially like to thank him for. John Urry drafted chapter 8 and contributed to the whole project through his Visiting Professorship at Roskilde University.

Material and findings presented in this book have partially been presented in other contexts. Parts of chapters 5 and 6 draw on a paper published in *Tourist Studies* (Haldrup and Larsen, 2003) and Jonas Larsen's PhD thesis *Performing Tourist Photography*. An extended version of chapter 7 has been published in *Tourism Geography* (Haldrup, 2004).

A number of people have been actively engaged in the project. First of all, Professor Wolfgang Framke, the director of the Tourism Research Centre of Denmark, managed the project from the beginning, while Jørgen Ole Bærenholdt took over the academic part of the management in 2001. Wolfgang Framke, Senior Researcher Per-Åke Nilsson (Centre for Regional and Tourism Research) and PhD student Flemming Sørensen (Department of Social Sciences and Business Economics) contributed significantly with case studies on the business side of destination development in Roskilde, Bornholm and Jammerbugten (Northern Jutland) that are reported in separate working papers. Wolfgang Framke, with his long experience in tourism research, also contributed to the production of this book at numerous seminars and meetings. Furthermore, Dr Elin Sundgaard, the academic secretary of the Tourism Research Centre of Denmark, provided important technical and social support to our work.

Drafts for chapters in the book have been presented at a number of scientific seminars and conferences in Lancaster (UK), Belfast (Northern Ireland), Rønne and Roskilde (Denmark), Höfn (Iceland), Gothenburg (Sweden), Vasa and Rovaniemi (Finland) and Alta (Norway). We are grateful for comments arising from such occasions, which have contributed to the networks and performances of this project. In particular, we would like to mention Kirsten Simonsen, Lars Aronsson, David Crouch, Hayden Lorimer, Harri Veivo, Judith Adler and Keld Buciek.

Skagen Museum, Bornholm Art Museum, Colberg Publishers and Lars Gornitzka gave permission to reproduce paintings and postcards in the book. In the final preparation of the book, in addition to Ashgate, we would like to thank the professional assistance of Joann Bowker, Claire O'Donnell, James Manley, Raymond Robinson, Poul Erik Nikander Frandsen, Rikke Larsen, Ingrid Jensen and Jytte Bach. Last but not least a large number of people interviewed have contributed to this book. The networks we produced with them, and their willingness to talk to us, discuss with us, and send us their holiday photographs and diaries have been absolutely crucial to our results. Without these networks there would have been nothing to report.

There are indeed many actors and networks involved in the complexities of research and book production; many not mentioned here. Much of these have been performed in everyday work within the Geography Section at Roskilde University and during workshops and seminars. There are both hard and soft infrastructures, from computer networks to the sociality of friends and families. And it is worth stressing that it would have been much harder, if not impossible, to produce this book without the fun and pleasure performed in connection with numerous face-to-face meetings. Therefore, the mobilities and proximities studied in this book are not unlike the practices involved in the very work of studying how tourist places are performed.

Jørgen Ole Bærenholdt
Michael Haldrup
Jonas Larsen
John Urry

Roskilde and Lancaster, August 2003

Chapter 1

Castles in the Sand

Figure 1.1 Tourist's photo: The sandcastle

Prelude on the Beach

If places did not exist the tourism industry would have had to invent them. Or if places did not exist the tourists would have had to invent them. Places are intrinsic for any kind of tourism. Without places to which to go tourism would seem meaningless. Indeed most tourism theories have revolved around the central theme: why do people go to places other than home for the sake of pleasure. Viewed as a central social and cultural practice in modern societies, tourism may seem simply a question of 'going places' (Feifer, 1985) or 'consuming places' (Urry, 1995). Tourism needs to 'take place' in order to work.

But although tourism obviously takes place through encounters with distinct places and place images, this book argues that tourist places should not be seen as necessarily involving strange, remote or exotic places. Instead following de Botton (2002) we argue that tourism is not so much about going places as it is about particular modes of relating to the world in contemporary cultures. Tourism is a way of being in the world, encountering, looking at it and making sense. It incorporates mindsets and performances that transform places of the humdrum and ordinary into the apparently spectacular and exotic.

Hence, tourist places are not bound to specific environments or place images. Rather it is the corporeal and social *performances* of tourists that make places 'touristic'. We stress how the human body *engages* with the natural world and hence produces spaces and places, rather than simply being located within them, or having them inscribed on its surface. The more traditional geographies of such inscriptions and markings are of little interest if they are not seen as human practices of production and remembering. As Crang remarks: 'analysing inscription and marking, without looking for practices, can only produce a mortuary geography drained of the actual life that inhabits these places' (1999, p.248). The human body is able not only of making particular anticipations and intuitions about the world but also of preserving the embodied memories of it.

Central to this understanding is a view of tourism performance that does not view it as isolated 'islands' of pleasure, theatre and so forth. Instead we avoid all sorts of purification (Latour, 1993) and instead stress the need to understand tourist places as hybrids bridging the realms of humans and nonhumans. This bridging is brought about by diverse mobilities and proximities, flows of anticipations, performances and memory as well as extensive social-material networks stabilising the sedimentated practices that make tourist places. We can illustrate this by turning to *the* emblematic tourist place, the sandcastle.

Like sandcastles, tourist places are tangible yet fragile constructions, hybrids of mind and matter, imagination and presence. The castle only comes into existence by drawing together particular objects, mobilities and proximities. The fine grains of sand with its alternating wet and dry textures, seemingly dead objects for decoration and stabilisation such as mussels, dead fish, stones and sticks, are drawn together by eager children's hands and helpful parents fetching buckets of sea water for the moat. Only then the pride of the family in the form of a castle of sand comes to life, towering over the beach. By building the monument of the castle the anticipations and traces of future memories are materialised on the beach, bringing together memory flows, objects and matter in making a hybrid tourist place.

The 'place', the sandcastle, however only appears as artefact as it organises a multiplicity of intersecting mobilities. There are *mobile objects* such as the fishes, stones and mussels found at the shore or on the beach as well as the tools spades, shovels, buckets brought in the trunk of the family car, possibly imported goods made in China or a similar low-wage country. The construction work also intertwines with different kind of *corporeal mobilities* (the day trip of the family, the journey to a holiday region, dense choreographies of construction work) as well as *imaginative mobilities* (long winter nights spent dreaming of summer beaches).

However these mobilities implies particular forms of proximity. The very making of a sandcastle is a social project. It requires face-to-face proximity among the family members constructing it and/or acting as an impressed audience. It also relies on a face-to-place proximity with certain objects, scenery and landscapes.

The sandcastle, the sea, the radiant sun are set-pieces for staging a particular performance, a carefree afternoon at the beach. They become setpieces and backcloth for staging a moment of pleasure worth remembering. But they are not pregiven. The scene is only produced, the set-pieces and backcloth drawn together and inscribed with particular meanings, when performed. The very making of the sandcastle is the realization of such a moment of remembrance. Through the sand castle, the space of the beach is domesticated, occupied, inhabited. The castle transforms the endless dull masses of white, golden, fine grained or gravelled sand into a habitat, a kingdom imbued with dreams, hopes and prides. This transformation of the beach into a social space is unthinkable without the vast and extensive networks stabilising and reproducing it as a particular tourist place: the network of roads necessary for accessing the beach; holiday housing, camping sites, beach hotels, restaurant and bar districts facilitating the visitors; commercial marketing and internet advertisements feeding the imagination of potential visitors; public holiday acts and legislation on housing constructions zones; widespread ownership of private cars; access to road maps and guidebooks; and the institution of the nuclear family. All these are parts of networks that stabilise and regulate the sedimentated practices that in the end makes that sandcastle on that beach.

What we say of the sandcastle here can be generalised. Other tourist places rely on such a transformation of landscapes, sites, attractions, cities, buildings and so on into social spaces. Neither the material existence of a physical place nor the memory of particular pleasurable visions makes tourist places come into being. These are nothing but potentials, possibilities, dreams, anticipations. Places however only emerge as 'tourist places' when they are appropriated, used and made part of the living memory and accumulated life narratives of people performing tourism, and these performances include embodied and social practices and traces of anticipated memories. Tourist places produce particular temporalities. They are inscribed in circles of anticipation, performance and remembrance.

Likewise with sandcastles. Anticipated by expectant and impatient children, constructed with engagement and eagerness the castle rises as the masterpiece, the high spot of the day. For a couple of hours the castle is centre stage for the performance of play, and the applause of an admiring audience. It is the centre for this happy moment of pleasure and joy. As the afternoon arrives the sea rises and slowly erodes the fortifications. The family leaves. Waves role gently on the shore and at the end of the day no trace of the performance of the day is left. All is washed away and the castle only towers in the memory of the family, on the celluloid pictures brought home, and in the anticipation of the next day on the beach.

Envisioning Tourist Places

Tourism has become a significant aspect of contemporary social life. International travel in and out of Western Europe, Japan, China and North America particularly reveals substantial growth rates. To a degree not experienced before, the remote and the nearby are woven together in webs of mobilities. Hence, lives are increasingly defined by hybridities of home and away; hybridities not only produced by the emergence of the far-away in the midst of daily lives but also in many ways places are used, appropriated and domesticated for tourist purposes. While substantial social and cultural research have been carried out on the 'globalization of the tourist gaze' (see Urry, 2002, p.141ff) little sustained social and cultural research have been undertaken on more laid-back forms of tourism.

The book is about how people go about making places as part of our everyday lives by tourist performances. Hence it treats tourism as a way of encountering and sensing the world, and not as a specialised activity designated to, and appropriate for, very particular places. The places examined in this book are not awesome or spectacular places, not the Taj or Machu Picchu.

But the places we examine here are still visited and valued for the 'pretty ruins', 'nice views', 'tranquillity' and 'good times' they offer. The places only emerge as tourist places when viewed with particular mindsets and performances that transforms the humdrum and ordinary into places of excitement and extraordinariness. Places acquire their connotations only when made part of the broader social performance of the family holiday. While the chapters in this book highlights an often neglected aspect of contemporary tourism, the summer vacation spent in rented second-homes, conclusions of this book carries significant lessons for the study of other tourisms. As Franklin argues the deliberate attempt to slow down time found in much second-home based tourism may be as a significant element of contemporary tourism cultures as the spectacular emergence of the seaside resorts in the late 19th century (2003).

Hence, this book seeks to re-conceptualise tourism as a social and cultural practice. By contrast with prevailing understandings of tourism we understand tourist places as neither produced by the tourist industry and planners nor by tourists. Even thoroughly designed tourist bubbles such as theme parks and integrated holiday centres includes a materiality that escapes design. However, they only become meaningful tourist places through the processes of production, whereby human engage with the world. The subject of this book is how tourist places are produced, consumed, encountered and made sense of. In short, we examine how tourist places are performed and produced.

This approach differs from conventional approaches in tourism studies. In economic and environmental studies of tourism places has been understood as locations, where tourism development, politics, and planning as well as 'tourism impacts' happen (Pearce, 1989; Hall, 1994). More than theoretical development of frames of understanding, the focus has been on applied research (Hall and Page 2002). In such research the concepts of place, space and time have not been much considered. The implicit understanding of places in this line of reasoning has been as territorially fixed entities. Hence, the agency of tourist travel is viewed as an external element that destinations have to cope with.

In contrast the social and cultural theory of tourism has focused on the discourses, mythologies and imaginations that frame how places are perceived (Shields, 1991; Gregory, 1994; Urry, 1995). The implicit understanding of place suffers from a 'hegemony of vision' that reduces places to the visual formations constituting place images. Hence, this neglects how places are sensed, used, and practised.

Few studies have exmained how tourist places are performed in practice. Consequently the cultural analysis of tourism has tended to remain within a 'vicious hermeneutic circle', abstracting from how people relate to tourist places in practice (Squire, 1994; Crang, 1997). This critique of cultural analyses has drawn inspiration from the upsurge with social theory of the 1990s in examining the role of embodiment. It has been argued that tourism theory has neglected the role of the sensing, gendered body (Veijola and Jokinnen, 1994) and non-visual performances (Perkins and Thorns, 2001).

Such critical interventions and critiques have paved the way for a third position in tourist studies. This approach partly continues the semiotic insights of cultural and theoretical work of tourism (MacCannell, 1999; Urry, 2002; Rojek, 1993), but stresses more broadly how tourist practices are performed and intersect with everyday life in (post)modern cultures. Here the actual 'doings' of tourists in their complex relations with everyday life form the focal point (Franklin and Crang, 2001; Edensor, 2001). This attention to the practices and performances of tourists can be seen as a turn towards the embodiment and sociality of practices and places, also found elsewhere in contemporary social and cultural research. Instead of conceiving places as either fixed locations or cultural imaginations the making of places through performance is seen as central. Coleman and Crang argue. 'Instead of seeing places as relatively fixed entities, to be juxtaposed in analytical terms with more dynamic flows of tourists, images and culture, (...) we need to see them as fluid and created through performance' (Coleman and Crang, 2002, p.1).

This turn towards tourism performances is marked in recent studies (see Edensor, 1998; Desmond, 1999; Coleman and Crang, 2002b; Franklin, 2003). The performance and staging of tourist sites and events by the workers and managers in the tourism industry is evidently a significant feature. Moreover tourism (and vacation) can be viewed: 'a cultural laboratory where people have been able to experiment with new aspects of their identities, their social relations, or their interaction with nature and also use the important cultural skills of daydreaming and mindtravelling' (Löfgren, 1999, p.7). While the metaphor of the theatre thus seems an appropriate way of grasping tourism performances it also carries some problematic connotations regarding space. 'Too often', Coleman and Crang points out, 'dramaturgical metaphors suggest performance occurs in a place – reduced to a fixed, if ambient container' (2002a, p.10). Hence, the multiple ways in which the 'stage' for tourist performance is socially produced is not sufficiently examined.

In this book we take this 'performative turn' as our point of departure. However, there is a need to reconsider the spatialities and temporalities of tourist performance in order to examine both place and performance. As Franklin argues, the observation that the tourist world is increasingly indistinguishable from the everyday world does not account for the substantial ritual and performative nature of tourism (2003). It does not address the forms of intersection, nor how they relate to the apparent shifts in typical tourism practices. In the next section we elaborate on how spatialities and temporalities in tourism are involved as intrinsic elements of its performance.

Sedimentation and the Work of Erosion

Earlier we invoked the sandcastle as a metaphor for capturing significant characteristics of tourist places. We emphasised how sandcastles could be viewed as hybrid artefacts, drawing together mobilities and proximities that crosscut the realms of the social and the material. We discussed the different proximities such places imply and showed how a sandcastle affords a dramaturgical landscape for staging the roles and bonds of the people building it. Finally, we discussed the extensive networks that stabilise and regulates the sandcastle and the beach as a tourist place. In short we used the metaphor of the sandcastle to highlight that the production of tourist places relies on drawing together particular mobilities and proximities to afford a stage for performances. When we here use the notion of 'production' we do so in the broadest possible sense in order to capture *all* sorts of social practices. Hence, we understood space as a socially produced *spatiality* (Lefebvre, 1991; Simonsen, 1996). In doing this we depart from the idea of fixed space in order to replace it with 'an image of a complex of mobilities, a nexus of in and out circuits' of objects and energy, information and images (Lefebvre, 1991 p.93; see Urry 2003a, pp.48-50).

Here we can turn again to the sandcastle that is fluid and changing. Tunnels and towers may collapse as the sun shines, the wet sand dries up, and the texture changes. The rising tide may cause water to penetrate the ramparts surrounding the moat and undermine the fortifications of the castle. The work of erosion and sedimentation alters the sandcastle slowly or with sudden ruptures. It changes their appearance and inspires new reconstructions, a submerged moat may inspire the construction of a channel to the sea, a collapsed tower affords space for an enlarging the stronghold.

The work of erosion and sedimentation also applies to other tourist places that may be eroded, overlain by new sedimented practices and/or reconstructed at a later time. The experience of many beach resorts and hotels of the late 19th and early 20th century come to mind. The rise and fall of seaside tourism in Northern Europe mean an erosion of the former shrines of beach tourism. Hence, the sedimented practices of the 19th century seaside resorts were eroded and overlain by new modes of sedimented practices (see Shields, 1991 pp.73-116; Urry, 2002, pp.16-37). Only later did these processes of erosion and sedimentation inspire reconstruction of such places. Hence, in the late-20th century early seaside resorts had been reconstructed as a centre of gambling (Atlantic City) or as a stage for 'slowing down time' as in Scandinavian historic beach resorts.

The idea of sedimentation can be further examined through Ingold's argument that much social theory has assumed a false antagonism between humans and the environment. Instead of positioning humans over and against 'reality' he suggests a shift of perspective. Drawing on Merleau-Ponty he contends that the human body is not so much in space but belongs to space. Bodily practices are already oriented towards actions in the world (Merleau-Ponty, 1962, p.142). Hence, humans are inscribed in the world and do not merely ascribe meaning to it. They inhabit it from their birth onwards, they use it and their capabilities (language, tool use and so forth) are products of this active use. This *dwelling* perspective, as Ingold coins it departs from a conventional Cartesian ontology and this has profound implications for how to conceive of place (2000, pp.185-87).

By contrast with approaches that conceive the cultural construction of place as a process of inscription, Ingold views this as a process of sedimentation where human embodiment, sociality and memory came to be incorporated *into* landscapes and places. Sedimentation is not about depositing different layers of symbolic meaning. Sedimentation may imply the essentialist idea that the true landscape can be found, by removing the symbolic blankets (Ingold, 2000, p.208). By contrast Ingold proposes a material constructionism in which material, social and cultural aspects of place sedimentation are integrated. This integration he captures in the two concepts of landscape and 'taskscape'. The notion of taskscape is pivotal as it refers to the ways humans inscribe themselves in space, by using, inhabiting and moving through it. 'Just as the landscape is an array of related features so – by analogy – the taskscape is an array of related activities' (Ingold, 2000, p.195). Similarly, he says: that: '*landscape as a whole must likewise be understood as the taskscape in its embodied form*: a pattern of activities "collapsed" into an array of features' (Ingold, 2000, p.198, his italics).

Both landscapes and taskscape are concepts that point to the fundamental temporality and sedimentation of spatial practices. Temporality is not only relevant to deeply sedimented practices and symbolic forms that we may encounter when seeing harvesters at work in a field, but also when people engage in this landscape, taking possession and dwelling within it. It is the tracks and paths that crosscut the landscape that makes the taskscape visible. In that sense: 'there can be no places without paths along which people arrive and depart; and no paths without places that constitute their destinations and points of departure' (Ingold, 2000, p.204).

Hence, only as people 'feel their way' through the world that it comes into being. Ingold continues: 'Just as with musical performance, wayfinding has an essential temporal character (...): the path like the musical melody, unfolds over time rather than across space. (...) In music, a melodic phrase is not just a sequence of discrete tones; what counts is the rising or falling of pitch that gives shape to the phrase as a whole. Likewise in wayfinding, the path is specified not as a sequence of point-indexical images, but as the coming-into-sight and passing-out-of-sight of variously contoured and textured surfaces' (Ingold, 2000, pp.238-39).

Temporality is fundamental to the immediate experience of places and landscapes, as well as to the many deeply sedimented practices inscribed in them over time.

Tourist Times, Tourist Spaces

The metaphors of sedimentation and erosion indicate that places may be fluid, but they are fluid in specific ways (Crang, 2002). Important aspects of such 'sedimentations of place' involve how people and things are transported to, from and through places. Places are assigned particular roles in the performance of tourism. Memories are reworked by sharing memories with holiday companions or friends and family at 'home' or through the imaginative mobilities bridging the sequences before, while on and after the holiday.

Tourist places are produced, not only by the actual performance of tourists as much of the tourism performance literature suggests, but also by the stabilising and intersecting flows of people, objects, memories and images. This attention to networks and flows, bridging the dualisms of home and away, the cultural and material, physical space and social space, raises the question as to how to capture the mobile or fluid production of such tourist places.

One inspiration is here the time geography. It focuses on the 'interpenetration of technology, society and landscape' (Hägerstrand, 1983, p.250), bringing out its implications for the spatio-temporal choreography of social life (Ellegaard 1999; Lenntorp, 1999; Thrift, 1996; Hägerstrand, 1985; and on tourism, see Aronsson, 1997). However, by translating individual paths into an abstract, Euclidian time-space, landscapes and objects problematically emerge as a fixed and neutral background for people's trajectories through time and space. 'If corporeality is taken to its own limits,' Gren argues 'it also implies that we are in fact dealing with *multiple* corporealities' (2001, p.212). The modalities of movement cannot be reduced to a graphical trail. The 'geographical map' and the 'history book' are complementary elements of the practice of movement (de Certeau, 1984, p.98, p.120ff). The spatio-temporal paths of people can (and should be) plotted against the multiple time-geographies produced *by* moving bodies.

As de Certeau contends 'the foggy geography' of the city is constituted by the performance of walking in the same way as speech acts constitute a language (1984, p.104). Following this we may argue that different sorts of mobility (walking, biking, driving and so forth) are constitutive of places (Macnaghten and Urry, 1998, p.204). We may further argue that temporality ma be a key aspect neglected both in tourist studies and theories. Hence, deliberative attempts to stop time, stepping outside of the everyday time flows, experiencing what Lash and Urry have called 'glacial time' are crucial aspects of contemporary tourism (1994, p.241). Diverse tourist practices such as those related to second-home tourism, hiking and trekking and the upsurge in demand for extreme sports tourism, can all be said to contain aspects of this replacement of instantaneous and clock time by a more inert, slower glacial time. Franklin proposes that this temporal aspect of tourism is related to shifts in how time is experienced in contemporary cultures. It is precisely the contemporary 'tyranny of the moment' that is decisive for this emergence of 'slow-time tourism' (2003). Franklin goes on to predict that deliberative aspects of slowing down time while having time off will be even more significant in the future (and see Eriksen, 2001).

However, the fluidity of places is not only a question of corporeal mobility but also of mobile objects and imaginative mobilities. The temporalities and spatialities of tourist practices are simultaneously material, social and cultural. They are material since they involve the movement *and* the fixing of bodies and things, social since they involve social interaction and networking with others, and cultural because of the role of collective imaginations. These mobilities interweave in the circuits of anticipation, performance and remembrance that characterise tourist practices. Tourist practices are inscribed in the circular sequences before, during and after the 'travel' itself.

In his seductive essays on the art of travel, de Botton describes how the overexposed photographs of palm trees, clear skies and white beaches are set in motion on a lengthy and expensive journey to Barbados (2002, pp.8-9). After having arrived to his destination he walks to the beach depicted in the brochure, recognises it, and immediately his mind starts making its own wanderings: from the troubles and physical pains caused by the flight, the prices of meals and eventually his mind migrates to visit a scheduled work project for next year. De Botton concludes: 'There is a purity both in the remembered and the anticipated visions of a place: in each instance it is the place itself that is allowed to stand out' (2002, p.22). He even adds '...it seems we may best be able to inhabit a place when we are not faced with the additional challenge of having to be there' (de Botton, 2002, p.23).

The imaginative mobility before and after the holiday intersects with the inert, troublesome, and sometimes even hazardous practices of tourism. Hence, the times and spaces of tourism are not limited to the particular tourism regions and sites but also comprises the dreamscapes of anticipation and remembrance, as well as different social and material networks crosscutting bounded spaces. We can elaborate these sequences by outlining the various mobilities involved here.

The imaginative mobility performed when not on holiday involves dreams and anticipations. These dreamscapes are fed by the circulation of images, travel narratives, souvenirs from earlier vacations in the same or adjacent places or bought by friends and relatives, and also the tourism marketing industry. Spatial representations materialised in brochures, guide books or on the Internet provides detailed accounts of the many possibilities offered by competing destinations and supports detailed speculations and planning on whether to bike, walk or drive and which places to stay while on vacation. They can be explored and compared with representations gained from other sources, such as school lessons, the Discovery channel or fantasy literature. Tourism is closely related to the shaping of lay geographies people use to make sense of their world (Crouch et al, 2001). These lay geographies depend on extensive networks of objects, institutions and sedimented practices that regulate and stabilise the expectations and anticipations.

We have dealt with some networks and mobilities that stabilise the performance of tourist places. We argue that 'tourist places' are produced not only by the meanings ascribed to them by individuals and/or groups. They are also constituted by different corporeal mobilities. These sedimentated mobilities afford the constitution of different tourist places (see Edensor, 1998, on different tactics of walking). Hence, a sightseeing trip involves the intersecting practices of gazing and glancing at sights and passing landscapes (Larsen, 2001). Although the visual sense clearly dominates this particular mode, it does only include a disembodied and shallow experience. Sightseeing does not exclude aural and haptic sensations and corporeal engagements in the world. Other ways of engaging in the world while on vacation involve different configurations of the senses. Hiking, mountain bikes and different sorts of playful activities such as parachuting and mountain climbing more directly emphasise haptic sensations (see studies reported in Macnaghten and Urry, 2001).

A further aspect is the social character of tourism. Hence, the aspect of simply 'being together', whether with peers, colleagues or as in this book family and friends. They can be crucial for many sorts of travel and are often highly entangled with how sites and landscapes are encountered and sensed.

A further component of the tourism performance is what happens after the return. Memories from travelling materialise in photo-albums, souvenirs exposed at home and given away. Whether as means of showing off and making social distinctions or as private records of a communal life of friends and families, the materialised memories are important means through which 'tourist dreams colonize all those other fifty weeks, when we are not on vacation' (Löfgren, 1999, p.7).

Across the sequences of before, on and after vacation, threads of meaning are interwoven. Specific meanings are attributed to the places visited; meanings that incorporates sites and landscapes into the living experience of the people visiting them. Places are not only or even primarily visited for their immanent attributes but also and more centrally to be woven into the webs of stories and narratives people produce when they sustain and construct their social identities. Hence, tourist places are hybrid places of home *and* away. These hybrid places are not only played and placed in limited timespace but overlap with each other, drawing mobile objects and imaginative mobilities into the cycle of tourism performances.

Performing Tourist Places

Tourist places are produced spaces and tourists are co-producers of such places. While tourist places therefore can take many different forms, and the same environments may be encoded in very diverse ways, the performances constitutive of tourist places are sedimentated practices requiring extensive networks and flows of mobilities in order to stabilise. This book deals with the multiple types of networks and flows of mobilities involved in stabilising and regulating the performance of tourist places.

Most environments attractive to tourists have not been produced for that purpose and they have other histories and geographies of nature, society and culture. There are institutionalised patterns and narratives of tourism that tourists more or less follow or react to in the production and performance of tourist places. Experiences, patterns and narratives are communicated through 'networked mobility' that are social and technical relations moving information and thereby contributing to the lay geographies of potential visitors. Ideas of what one can do where and when circulate, so that tourist imagination and expectation can prosper. Networks mediate physical transport, social relations and cultural imaginations, and they do so before, during and after specific tours to a number of specific places. In that way the production of tourist places is based on the intersection of professional and private networks. Professional representational strategies in marketing do of course contribute to the making of tourist markets, and thereby also to tourists' expectations. But the market cannot be taken for granted. The market has to be produced and reproduced in certain ways as an institutionalised and regularised way of transaction. Hence, the flows of tourists to and from the region, as well as goods, touristic objects and representations, construction of facilities, housing, running of tourist oriented shops and so forth is in important ways managed by local tourist organizations and business networks.

Chapter 2 deals with the networks that are producing tourist places, asking whether these business networks constitute an 'industry' and how they stabilise tourist places. Chapter 3 moves the focus to the consumption practices performed within particular places. While chapter 4 shows how different tourist practices stage the particular site of the beach in distinct ways, chapter 5 examines the visual practices people employ when photographing at tourist sights. Both chapters discuss how 19th century artworks and guidebooks contributed to the scripting and staging of current tourism spaces and practices. In chapter 6 we show how people take photographs to produce memory-stories and chapter 7 examines the various modalities of tourist mobility. Finally, chapter 8 sets the agenda for a 'new mobility' paradigm in tourism studies, discussing place, proximities and the mobile methods needed for further research.

Chapter 2

Producing Tourist Places[1]

It is evident from the behavior of global corporations during the last thirty years that they are being driven half mad as they circle the human relations at the heart of the world's largest "industry"; that is, as they attempt to come to terms with the fact that the economics of sight-seeing is ultimately dependent on a non-economic relation (MacCannel, 1999, p.196).

'Destination' Policies

Places possess certain material and imagination-stimulating qualities. They are able to attract tourists because they also represent these qualities *prior* to the arrival of tourists. Among these qualities, the supply of transport, accommodation and food to meet tourist demand is important. While the following chapters look more into the performances of tourists, this chapter investigates how tourist places are produced through policies, industries, and by building facilities.

In the discourse of tourist policies and tourism management, tourist places are normally treated as 'destinations'. In addition to attractions, 'destinations' also embrace tourist facilities in the form of transport, accommodation and food. The concept of 'destination' is used for place marketing, where the contributions of businesses, museums and other attractions are combined in brochures and websites by tourist agencies. 'Destination' is also used to model ideal local or regional networks, thus creating a total product of services and experiences to meet tourist demand (Framke, 2002a).

Not much tourism literature has analyzed such networks in destinations. They have usually been approached as a question of organization ('destination organizations') with marketing as a major task (Pearce, 1992). Yet 'destination' is a dominant concept or metaphor in much tourism policy and applied research. But there has also been critical discussion of whether the approach really helps policy and research (Leiper, 2000).

---

[1] We would like to thank Wolfgang Framke for his assistance in writing this chapter. Each of the three areas researched here has been investigated in a separate paper, reporting on a range of qualitative interviews with destination managers, local developers and planners, and representatives of tourist facilities and attractions in Jammerbugten (Sørensen, 2002a) and Roskilde (Framke, 2002b). The paper on Bornholm (Nilsson, 2002) was based on a range of earlier investigations. The research was presented and discussed at small conferences for tourism operators in each of the three areas.

The idea that businesses involved in tourism should cooperate in terms of 'destinations' is expressed in national and regional tourism development policies in various countries. The orthodoxy of 'localized learning' is implemented in policies of 'learning destinations'. Linkage of innovations and destinations has been an organizing principle in the tourism policies of the Denmark government.

In the Danish Government's Tourism Policy Reports, which have been published since 1986 (*Industriministeriet*, 1986 and 1991; *Ministeriet for Kommunikation og Turisme*, 1994; *Erhvervsministeriet*, 2000 and 2001; *Økonomi- og Erhvervsministeriet*, 2002), cooperation among private tourism enterprises as well as with public authorities has been in a general focus. Reports have resulted in subsidies for inter-firm cooperative action in product and destination development. National level support is also given to regionally organized Tourism Corporations for coordination, marketing and strategy development.

In 2000 the construction of consortia was suggested in order to facilitate product development, and the 2002 Report suggested national 'tourism alliances' within specific 'business areas'. The destination perspective has been developed in a trend from the local towards the national. So while the idea of local processes of destination cooperation was stronger in the early 1990s, interest is increasingly concentrating on the branding of Danish core values (*Erhvervsministeriet*, 2000) and national growth (*Økonomi- og Erhvervsministeriet*, 2002). In reality, this national policy discourse is abandoning destinations at local/regional levels, and builds on the assumption that tourists visit Denmark because of its 'national' qualities.

However, the amalgamation of tourist organizations in the whole of the metro-politan area (including Roskilde and northern Zealand in the Wonderful Copenhagen organization) from 2003 is accompanied by discourses on the development of 'an effective concept of destination'. But this is the discourse of the destination organizations' own internal need to rationalize management, staff and budgets in order to develop a 'common regional business system' (Nordtour et al., 2002).

The problem with the discourses of destination organizations, especially the 'professional' and large-scale ones, is that they tend to confuse the terminology of internal organization with that of inter-firm and other inter-organizational relations. Thus destination organizations produce an image of all-inclusive business activities, while much of this work is in fact undertaken independently of the management of such destination organizations.

This chapter questions the theoretical assumptions in such policy discourses. Informed by empirical studies in three 'destination' areas of Denmark, it examines the types of networking that organize 'destinations'. The three 'destination' areas are Jammerbugten, Roskilde and Bornholm (see Figure 2.1). The research findings here challenge economic geography's conventional understanding of territorially flexible specialization and localized learning. Programmatic ambitions emphasize the 'missed opportunities' of economic geography in tourism, as tourism has in the past been ignored in economic geography (Debbage and Daniels, 1998). But we will see why traditional economic geography approaches to place-bound networks do not explain the dynamics of tourism. The later part of the chapter suggests an alternative understanding of how tourist networking produces tourist places. This approach is used to understand how tourist places are generated in the case of the major tourist accommodation facility in Denmark, the 220,000 'second homes' or holiday homes. Finally, the concept of 'destinations' itself is deconstructed.

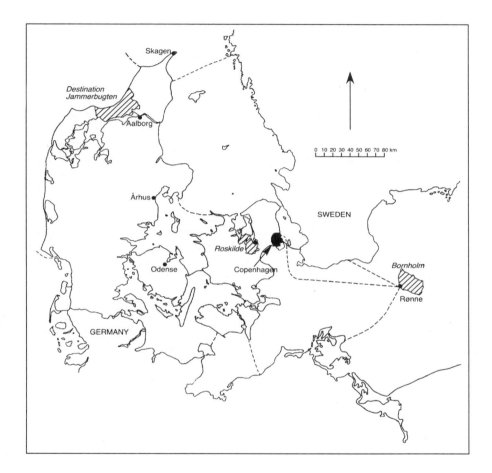

Figure 2.1   Map of Denmark with destination areas Jammerbugten, Roskilde, and Bornholm

The 'Tourism Industry'

Traditionally, economics, economic sociology and economic geography approach industries as producers of certain goods and services, and specific industries are identified by the very products they produce. Thus if tourist experience can be seen as one product, the tourism industry would be expected to be *one* industry. This would mean that businesses in *the* tourism industry would have potential common agendas for cooperation (and competition), because they produce the same kinds of products. The debate between Smith and Leiper on the question of whether tourism is *one* industry is thus illustrative (Smith, S.L.J., 1988, 1991, 1993; Leiper, 1990a, 1993).

Smith argued that a supply-side definition of tourism makes it possible to define tourism as one industry. Leiper did not accept Smith's supply-side definition, since tourism industries also depend on the demands and behaviour of tourists. 'Most organizations that supply goods and services to tourists (...) are *not* in the business of tourism. They might have tourists amongst their customers' (Leiper, 1990b, p.149). Therefore, understanding tourism as a single industry is inadequate.

In later reconsiderations, Smith now defines the tourism industry as something that at certain moments forms a political network. 'Although these businesses traditionally do not cooperate much with each other to promote their common interests, they sometimes rally together to express frustration over government policies, regulations, and taxation systems that affect them all' (Smith, S.L.J., 1998, p.32). The tourism industry is produced when political networks exercise power by coming together and using the discourse of 'the tourist industry'.

'The tourist industry' only comes into being when businesses with interests in tourism come together to express common interests related to their 'general conditions'. They typically address governments, and there is a tendency to 'blame the other(s)' rather than scrutinize their own business performance. This is clearly expressed by the director of the Tourism Corporation of Middle and Northern Jutland:

> ...for many years, large parts of the industry have argued politically, and have therefore come to see it as the definitive truth that when problems arise in our industry, they are not our own fault. They are because of high VAT rates, or the weather being so bad, or the fact that environmental taxes have suddenly grown so high, or local government will not properly pre-sort waste, or whatever. That is, there has been a tendency, and there still is a tendency, to see all troubles as caused for us from the outside... (interview, quoted from Sørensen, 2002a, p.17, our translation).

While central actors in tourist organizations express such frustrations about the behaviour of the industry, they do not really question the existence of *the* tourist industry that they are professionally engaged to develop. Yet, beyond holding meetings and declarations, actors in tourism do little to construct common conceptions of an industry, neither among the industries nor in governments. Investigations show that cooperation can work in very specific agendas such as an association in a specific sector (e.g. of hotels) or destination organizations. But apart from the fields of destination marketing or offers of specific package tours, firms in transport and hospitality have difficulty imagining themselves fully cooperating.

Tourism is neither one industry nor does it produce one product. Tourists combine very different forms of commercial and non-commercial facilities and attractions when they 'perform' tourist places. Even when combinations are packaged in tours, the experience does not become one 'total product' prior to the performances of the tourists through shopping, relaxing, photographing and so on. And this too is why Smith finally deconstructs the conventional understanding of the tourist product: 'The consumer is an integral part of the production of the tourism product. In fact, the consumer actually produces the final tourist product – the experience. Unless the consumer actually uses the intermediate outputs in the course of being a visitor, there is no experience, no final tourism product' (Smith, S.L.J., 1998, p.51). The process involves a variety of 'industries' including the self-producing tourists. Word-of-mouth marketing, too, is crucial to tourism, insofar as tourists themselves have to be acknowledged as crucial actors in the tourism business flows that transmit place information and images.

Given the crucial role of tourists within tourism, what is the significance if any of networking by businesses, organizations and authorities?

Economic Geographies of Tourism

The concept of network has been widely used in theories of inter-firm relations and different types of bonds (Håkansson, 1989). Networks are thought to support and implement transactions, solve technical and logistic problems, deal with administrative and juridical procedures and more generally facilitate learning and trust. It takes time to build such networks or bonds, but potentially they reduce costs. Networks are seen as alternatives to, or modifications of, markets. There are three ways of understanding how they work in relation to markets. First, networks can construct the structures in markets in general. A second variation approaches networks as organized markets. The third alternative view networks as an alternative in between markets and hierarchies, as argued widely in the literature of economic sociology, institutional economics and innovation theory (see Lundvall, 1993).

In the 1980s and 1990s, economic geography and regional science focused on the 'resurgence of regional economies'. In the light of industrial districts like 'the Third Italy' and Silicon Valley, localized networks were seen as forces generating post-Fordism, flexible specialization and learning economies (Storper, 1995). A general paradigm has been built up for localized learning and learning regions, but in reality much of the argument was based on specific cases of manufacturing industries, where tacit knowledge is crucial and develops within relations of proximity (Maskell et al., 1998).

Much of this literature implicitly assumes that non-economic factors such as social capital and 'culture' are territorially embedded, and that 'localities' and 'regions' are homogeneous, territorial actors or networks. In reality, such understandings must be questioned (Harloe, Pickvance and Urry, 1990; Bærenholdt and Aarsæther, 2002). However, inter-firm network theory has a point in stressing the importance of relations lasting over *time*. A network means that one can count on the support of others inside the network. There is cooperation through reciprocity over time, but *without* needing constant proximity.

Few have questioned the new orthodoxy. Amin and Thrift are critical of the assumptions of *localized* learning and the necessity of everyday proximity for sharing tacit knowledge (Amin, 1993; Chapter 3 in Amin/Thrift, 2002). What matters in agglomerations, such as cities, rather than inter-firm relations, is proximity to *consumers.* Furthermore, agglomerations are 'light institutions' facilitating the proximity of *meeting places.*

> Restaurants, bars, football matches, musical events, golf clubs are places where ideas are developed and deals are struck, deliberately, or through casual socialization. They are also places where standards are tracked, gossip is exchanged, rivals are noted, and disputes are aired, rather like in business associations and interest groups. But, these meeting places are not Marshallian spaces of interchange between members of the same community of interest (say, furniture makers in an Italian piazza). Instead, they are more broadly constituted centres of sociability or professional gathering with a light economic touch; mixing pleasure, voice, search and business opportunity in emergent ways (Amin and Thrift, 2002, p.73).

Such forms of networking do have economic functions, but they are also pleasurable sites for socializing, and thus stabilize economic functions. To these we can add the milieux represented by agglomerations of various educational and research organizations and of certain labour markets with people qualified for specific types of work.

While there are critical discussions of approaches to industrial districts and regional economics in general, the use of these understandings has been particularly debated in tourism research (Sørensen, 2002b). This finds that tourism is generally weak in networking, embeddedness, interest group representation and institutionalization (Williams and Shaw, 1998, p.376). Studies of networks in local tourism development show that networking has little effect (Tinsley and Lynch, 2001). Mackun compares networking in tourism in the Province of Rimini and networking in manufacturing in other parts of the Third Italy; it seems that the character of tourism and manufacturing networks are quite different (1998). Business networking and interactions in tourism are about future planning, marketing, and communication of collective needs to government (Mackun, 1998, p.268). Such forms of interaction are different from the industrial networks of inter-firm relations described in the Third Italy industrial literature.

Still, tourist destinations have been analysed in the industrial district framework (Hjalager, 2001). The logic seems to be that since tourists demand complementary products – such as transport, accommodation, food, sights and activities – cooperation and networking among such specialized producers are crucial. Scott considers tourism to be similar to other kinds of cultural production, although he specifies that in tourism 'consumers must travel to the point of production in order to partake of the immobile stock-in-trade' (2000, pp.205-6). Hjalager discusses the economic structures of tourism destinations with respect to the 'interdependence of firms', 'flexible firm boundaries', 'cooperative competition', 'trust in sustained collaboration', and 'a 'community culture' with supportive public policies' (2000, pp.201ff). She finds few of the features characteristic of industrial districts, and concludes that 'it is debatable how far tourism research can go using the industrial districts approach' (Hjalager, 2000, p.209). However, neither Hjalager nor Scott establishes the logical link between the absence of dynamic destination networking in tourism 'industries' and the performative and place-bound character of tourism. Instead tourism industries are blamed for not acting like 'industrial districts'.

Management-oriented tourism research has few illusions about destinations. Taking the market-driven character of tourism for granted, Poon analyses enhanced competition over the quality of services in the framework of vertical, horizontal and diagonal integration (1993). Since tourism industries focus on 'the delivery of finished services' (Poon, 1993, pp.216-17), vertical integration along the value chain is not as important as in manufacturing. Horizontal integration is more important, and it results in either concentration (in chains) or in diversity and therefore in a lack of cooperation. Hence, 'diagonal integration' is not governed by the logic of value chains. Diagonal relations involve organized contact with tourists to offer them very varied products, following economies of scope as a means of developing economies of scale. So diagonal integration is a concept of 'networking the tourist' and networking *with* the tourist – so that the same marketing and sales networks can be used for the supply of air travel, credit cards, special hotel prices, car rentals, insurance and so on. The connections among the different services offered are not organized around any specific tour. While the same customers are only approached with different products to achieve business synergies, diagonal integration is, in fact, a strategy of cross-marketing, and it is not defined in relation to 'destinations'.

Among the business networks important to businesses servicing tourists, there are many relations that are *not* organized in relation to territorial units such as 'destinations'. However, we investigated the array of networks involved in the production of 'destinations'.

Types of Networks and 'Destinations'

In the three 'destination' areas we researched, six different types of networking are identifiable. First, there are *vertical* networks based on the delivery of raw materials and services such as cleaning and washing. Such networks are often of a local character, although less so when they involve chains, but they have little to do with 'destinations'. Suppliers of raw materials and services are chosen on the basis of trust, reliability and of course price; but even when proximity is crucial to the meeting of these criteria, vertical networks are organized in spaces other than the tourist place of the 'destination'.

Second, there are *horizontal* networks and chains, for example between hotels. Here the theories of industrial districts and localized learning might be expected to work, but horizontal networks and chains in tourism do *not* seem to be organized according to territorial criteria.

Third, we find 'additive' networks in *marketing* managed by destination organizations at local and regional administrative levels. However, this kind of networking should be analysed as cases of marketing organization rather than networks of businesses, museums and so on, as they often only meet when destination managers actually bring them together.

Fourthly, there are *informal* social networks based on the relations among local people involved in tourism but who have their networks located in other arenas. Such relations are often much more important than professional networks, as they produce the trust found in 'non-traded interdependencies' (Storper, 1995). They are crosscutting networks, producing social capital among inhabitants in the same locality, when people cross sectors and professional fields by meeting in sports clubs, choirs and other voluntary organizations (Bærenholdt and Aarsæther, 2002).

The fifth type of networking is that of *political connections*, whereby businesses related to tourism express common interests in issues such as the municipal planning of local traffic or the municipal funding of destination organizations. This is the type of networking that produces tourism as 'an industry'.

Finally, there is *networking with tourists* that seems systematically underestimated in research and has not been sufficiently studied.

First then, we consider how these various networks intersect in *Destination Jammerbugten*, which is a coastal area of North Jutland defined by the borders of the four municipalities of Pandrup, Åbybro, Brovst and Fjerritslev (see Figure 2.1). Historically, tourism developed around 1900, and the coastal landscapes are the main attractions (see chapters 4 and 7 on its history and tourist practices). In addition, a theme park oriented towards children was developed in the late 20th century, but tourists accommodated in the destination area visit many attractions outside the destination area (in Aalborg or Skagen). Few tourists define 'Jammerbugten' as the single focus of their vacation. Holiday centres, camp-sites, and second/holiday homes are the major forms of accommodation.

Local cooperation among businesses is limited. Tourism development and marketing are a field for the major firms such as the national agencies for holiday house rentals, and in addition there is the regional-level Tourism Corporation of Middle and Northern Jutland. The research shows that there is hardly any connection between networking and 'Destination Jammerbugten'. Even the local business development officers in the two municipalities on the fringe of the destination area (Pandrup and Fjerritslev) have little connection with the destination area, and they are also considering the possibility of forming other destination organizations. Among business actors, the 'destinations' that are of interest to them differ. Some actors define their area of interest as larger, such as the whole of North Jutland, and they stress that foreign tourists have no sense of the meaning of 'Jammerbugten'. The theme park 'Fårup Sommerland' considers the 'destination' to be the much larger area where their visitors are accommodated. Other actors though, such as owners of holiday centres, perceive the local area as most important, or even perceive their own facility, a camping site or specific hotel, as the 'destination' *per se*.

While vertical and horizontal networking (the first and second type introduced above) are of no importance to 'Destination Jammerbugten', the third type of *additive* network is also problematic. There are several overlapping organizations for marketing, and the 'destination area' cannot be identified with any central place or attraction. The 'welcome centre' of the destination is located in the non-coastal municipality of Åbybro where tourists pass through but where almost no tourists are accommodated.

The limits of the 'additive' networking are described as follows by the day-to-day head of a Nature Visitor Centre:

People tend to focus more on competition than on cooperation [...] everybody pays the same, and then everybody also wants to get the same amount back. And I think, when looking at our advertisement in the Jammerbugten catalogue and in Middle-North's "From the Middle Up" [the brochure of the tourism corporation of Middle and North Jutland], then the risk is that it'll be toothless, because they don't dare write anything about anybody, since that would mean that others aren't mentioned. Either all 67 contributors are mentioned, or nobody is allowed to be mentioned (interview, quoted from Sørensen, 2002a, p.31, our translation).

Every single business actor tends to concentrate on the cost-benefit of efforts in relation to his/her own firm or organization. It is no coincidence that this image is presented by the leader of a visitor centre, as such organizations tends to have a more 'holistic' perspective on the destination and have less commercial criteria of success. The larger agencies for rentals of second/holiday homes use their own marketing channels, and the theme park cooperates with the Swedish and Norwegian shipping companies responsible for ferries to Jutland. 'Destination Jammerbugten' is a product of *(geo)political* connections among municipalities in an area where little formalized business cooperation exists. Hence, it does play a role that local actors know one other well. However, some of the most important networks are not organized in terms of the territory defined by the four municipalities. In addition to the existence of various marketscapes, the connections with revisiting tourists are also important (see the German family reported in chapter 7).

Second, then, *Roskilde* is a different kind of 'destination' comprising the municipalities of Roskilde, Bramsnæs, Ramsø and Lejre, with Roskilde as the centre (see Figure 2.1). The four main attractions are the cultural attractions of the Viking Ship Museum and Roskilde Cathedral in Roskilde, and Ledreborg Castle and the Experimental Archaeological Centre, in Lejre. But the major tourist event is the annual Roskilde Festival that attracts some 70-80,000 mainly young people from across northern Europe. In addition to these there are one-day sightseeing tourists from Copenhagen and Sweden, cruise ship tourists visiting by bus from Copenhagen Harbour, and car-based tourists passing through or staying briefly.

There are a number of hotels, a youth hostel and a camping site, but almost no second homes are rented out, and hotels are more oriented towards business tourists than towards holidaymakers. Few visitors to the cultural attractions spend the night in Roskilde, and among tourists who find accommodation in Roskilde the main attractions they visit are actually in Copenhagen. The destination is much influenced by its location in the metropolitan area of Copenhagen and is now to be integrated into 'Wonderful Copenhagen', the organization for Copenhagen viewed as a single destination.

Many actors in tourism organisations do not focus upon tourism since they just happen to have tourists among their customers. This reflects how the economic role of tourism in Roskilde is marginal compared with Jammerbugten or Bornholm. Many business actors do not perceive Roskilde as a destination. They tend to speak of destinations as markets (accommodation areas) consisting of potential users of their services or visitors to their attractions. Interestingly, they do not speak of local coope-ration and destination networks. The destination manager stresses that Destination

Roskilde would not exist if Roskilde were not perceived as part of the broader Destination Copenhagen. For example, the 'Copenhagen Card' also covers attractions in Roskilde. Apart from the destination brochure and website, the only cooperation is in the field of business tourism and these initiatives come from outside.

Local vertical networks do exist in the supply of raw materials and technical services, especially to firms in accommodation and restaurants, but many goods, services and workers come from all over Zealand and even Funen. Cleaning is typically outsourced. In the case of small firms, trust is important in such fields, while the price is more important in bigger firms. But such networks do not support any definition of Roskilde as a 'destination'; they only contribute to the development of the town as a centre for multiple trades and services.

The most significant network in relation to tourism emerges with the annual Roskilde Festival. The voluntary organizations arranging the festival have developed competence in event management, establishing a business which also arranges other events. The networks combine vertical networks of relations with firms supplying goods and a variety of technical services, as well as the important informal networks of voluntary organizations from an area that also includes neighbouring towns and Copenhagen. The spokesman for the Roskilde Festival explained:

> What Roskilde *does* have, or has developed, is that among the 17-18,000 volunteers there is a know-how no one else in Europe has. And that's why we have formed, and are in the process of forming [at the time of the interview in 2001] a production company allowing us to use this know-how for other purposes than the festival (interview, quoted from Framke, 2002b, p.27, our translation).

In measures that also draw on political connections, a number of business actors and, for example, Roskilde University have now been involved in various spin-off activities from the Roskilde Festival. There is a 'Musicon Valley' project for development of music businesses; the national Rock Museum will be located in Roskilde; and Roskilde University is involved with new programmes in the fields of communication and management. This is a political project that combines vertical, horizontal, political and informal networks in an explicit attempt to construct a certain kind of industrial district, but the relation to tourism is only marginal. And the networks involved are different from those performed informally in the sociability of meetings and parties among actors in tourism. Although 'Destination Roskilde' is involved, the activities related to the Roskilde Festival are developed in their own fields and networks, and these are not limited to the area of the destination.

The final destination is the island of *Bornholm* in the Baltic Sea. Tourism has here been an important part of economic, social and cultural life throughout the 20th century (see chapter 5). Today, people are mainly accommodated in holiday houses rented out through local agencies. Hotels, bed-and-breakfast/guest houses and camping also play important roles.

The isolation of Bornholm as an island is a crucial part of its attraction, and this makes it potentially different from the porous 'destinations' of Jammerbugten and Roskilde. Visitors travel to Bornholm by various ferry routes from Denmark, Sweden and Germany, and recently also Poland, in addition to more expensive air connections.

In 2000 the surface travel time to Bornholm was significantly reduced with the opening of the Øresund Bridge between Sweden and Denmark and the introduction of one of the world's fastest catamaran ferries between Ystad (Southern Sweden) and Rønne. It is now possible to reach Bornholm from Copenhagen in less than three hours, compared with at least five hours before. Visitor surveys document tendencies towards increasing numbers of tourist arrivals and of shorter visits. So far, Bornholm is still perceived and organized as 'one' destination.

But interestingly, networking in tourism on Bornholm is not much different from the more porous destinations of Jammerbugten and Roskilde. While few argue over the use of additive networks in marketing Bornholm, tourism businesses, like other sectors, are well known for their lack of cooperation. This is also true in terms of the difficulties of getting businesses to contribute to the marketing efforts of 'Destination Bornholm'. In addition to additive networks in marketing, the most important networks in Bornholm are informal networks and networks with tourists.

The inhabitants of the island know of one another, since they meet in a various contexts. They also have fairly stable networks with revisiting tourists (see chapter 3). For example, the social interaction between host and guest in bed-and-breakfast/guest houses on Bornholm is a vital network formed with tourists, to the extent that the discourse is significantly one of 'guests' and not 'tourists' (Kaiser and Andersen, 1996, p.131). Guesthouse owners say that first-time visitors to Bornholm may use 'Destination Bornholm' for information and booking but repeat visitors call directly by telephone when booking accommodation. The overflow booking is mainly offered to partners within an informal, local network. Informal networks also govern the activities and attractions that hosts recommend to their guests. Local informal networks and networks with tourism overlap in ways crucial to their steering of tourist mobility.

Many of the tourism facilities on Bornholm such as guesthouses, are owned by locals (Kaae, 1999), and tourism is much more important to the economy than in Roskilde. Thus many local people perceive tourism as a central aspect of their liveli-hood and their way of life. Moreover, the need for seasonal workers is greater in such areas. Tourism entrepreneurs and workers are often 'non-natives', originally themselves arriving as tourists. It is crucial to tourism that entrepreneurs can reproduce and develop their networks out-of-place, and the entrepreneurship of 'in-movers' is often due to their role as locally based networkers with tourists and outside partners. This has also been documented in a comparative analysis of case studies of networking in the development of tourism in Nordic peripheries in Greenland, Iceland, Sweden and Finland (Bærenholdt, 2002). Kneafsey's case studies of rural tourism development in Ireland and Brittany also show that networks lying beyond, but also involving local ones, were decisive in the successful development of tourism there (2000).

Our research shows that few dynamic business networks are associated with the organization and production of these three 'destinations'. While political connections and informal networks are more important among local networks, networks across place boundaries are also crucial to the development of tourist places. The implications of these findings for understanding networking in the production of tourist places are now discussed.

Networks Producing Tourist Places

Apart from the marketing efforts of destination organizations, the more innovative business networks in tourism do not fit the concept of 'destinations'. The networking that facilitates business innovation is different from the conception of destinations that is transmitted.

There is more integration in local businesses if some kind of destination package (a visitor's card or the like) has been bought. Still, this does not create a vertical or horizontal network among suppliers for the production of service and experiences. It is simply a kind of sales cooperation. If consumers buy different goods and services from different suppliers inside and outside the town where they reside, this does not necessarily make the relations among these suppliers an innovative milieu. The suppliers' cooperation and learning is not place-bound; it takes place in specialized sectors and organizations.

Hotels cooperate locally if they are fully booked and if they do not have partners in the same chain nearby. Most often, however, fierce competition limits cooperation, or hotels are merged into chains to obtain economies of scale. Individual hotels or different chains in the same locality seldom cooperate. Hotels can learn a lot from exchanging experience of their strategies in different places with other hotels, but these relations do not build on the co-location of the services. It is more likely that the opposite is the case. Hotels of a similar kind learn from one other by comparing the challenges associated with their own location with those in different localities. Furthermore, such networking can be more innovative and creative in other spaces of long-distance communication or meetings, where it is not embedded in the stress of everyday routines. Those involved may prefer the experience of meeting others in the meeting-places produced in business tourism.

In fact, the driving forces in destination organizations are not businesses but *local authorities* and attractions that may also have a public-sector character. Hotels, second-home agencies and car rental firms have other sites of learning and other marketscapes, and they do not need to orient themselves towards the attractions in the environment closest to their facilities. But when tourism development becomes a political strategy for securing jobs and revenue at the destination, marketing is the obvious first step. Most local authorities have websites with portals for locals as well as tourists. In addition to local economic development, destination marketing also has the political goal of enhancing the outside prestige of a locality in terms of political capital and power.

Local authorities (municipalities) play significant roles in tourism development because of their role in physical planning. In the historical development of British seaside resorts, the role of local government was already increasing in the late 19th century and became important to the improvement of sanitation, the construction of sea barriers and seaside promenades, and also to the maintenance of 'public order' (Walton, 1983, pp.125-29). Since the introduction of planning acts in the early 1970s in several countries, local authorities have been responsible for adopting binding local plans. Furthermore, the focus on sustainable development, and especially Agenda 21 initiatives, has further strengthened municipal involvement in local tourism development. Nowadays too, the spatial organization of the environment is a major tourism task that can conflict or converge with the interests of local citizens. Traffic regulation and the architecture of public space, including that of harbours, are important tasks of Danish municipalities.

In the case of Roskilde, local authorities are central in the recent reconstruction of the harbour area and the development of the Museum Island project as an attraction at the Viking Ship Museum in Roskilde. As an attempt to make visits to the Museum more colourful shipbuilding workshops have been established in a more interactive and open setting. To construct the site, local authorities raised some of the funding needed to match funding from various foundations. The project would not have been mounted without the political leadership of the Roskilde City Council, and the future of the attraction also depends on municipal action in financing and planning. The strategic visions of local government do not always work well in tourism (Cooper, 1997, p.91). But local authorities play a crucial role in local tourism development because of the immense need to coordinate business, infrastructure, and recreational facilities and landscape management.

Tourism networks are indeed material-spatial organizations of places and flows. Tourist places are produced in complex relations where the material production of place intersects with imaginative place production and this is organized through mobile and territorial networking practices. Various networking practices govern the practices of tourists, tourist businesses, tourist organizations, local authorities and other local people. Innovative horizontal networking among businesses and networking with tourists are mobile practices beyond the level of destinations, while practices in marketing, informal networks and political connections combine territorial and mobile networking to producing place images, cooperation in places and the materiality of tourist attractions and facilities in places. This combination of mobile and territorial networking is crucial. Tourist places are produced by the cooperative efforts of territorially defined relations *and* by the mobile interactions among tourist businesses, tourist organizations and tourists across place boundaries.

The production of tourist places involves dynamic networking practices. Various spatialities are embedded in the ways networking is practised. Space is not an external feature of networks (Bærenholdt and Aarsæther, 2002). Networks are practised; their potential stability depends on the practices producing them, and power relations are as porous as the networking that performs them. This understanding fits well with the way in which Actor-Network theory investigates the constant production and performance of markets as processes of networking (Callon, 1999; Dicken et al., 2001). Actor-Network Theory approaches the social and spatial production '*performances* that underlie processes of economic change and development'; it focuses upon 'the ability of actors to "act at a distance" by entraining both other actors and the necessary material objects, codes and procedural frameworks to effect the activation of power' (Dicken et al., 2001, p.102). Such an approach sheds light on the production of tourist places because of the complexities of the spatial patterns and the cross-cutting of material, social and cultural aspects.

Tourism research has been caught in territorial traps and linear metaphors of scale, as has much social theory (Urry, 2003a, p.122). It is a paradox that tourism industries, business networks and policies with their fundamentally mobile character have been researched through the prism of territorial categories such as 'destination'.

Furthermore, little tourism research has cut across the fields of materiality, the social and culture. Latour asks: 'Is it our fault if the networks are *simultaneously real, like nature, narrated, like discourse, and collective, like society?*' (1993, p.6). The networking that produces tourist places transcends distinctions among environment, sociality and discourse. Networks make their way through matter and bodies, and they are performed with conceptions, narratives, languages and identities. Networking is about the social and spatial relations of economies, policies and everyday life. It should be evident that 'production', 'producing' and 'producers' are understood in a broad sense, including much more than the 'material production of manufactured goods'.

Tourism is an organized, regularized form of mobility that works in complex networks producing places as material natures, social relations and discursive conceptions. The decentralized and contingent character of these networks can be seen in the organization of the second-home accommodation system in Denmark.

The Place of Second Homes

Almost 40 per cent of all tourist overnight stays in Denmark take place in second homes or holiday homes. The commercially rented 'second home' is the major form of accommodation in tourism in Denmark for foreigners, while many Danes stay in their own second homes or in the second homes of friends and relatives. An estimate from 1991 showed that there were almost as many non-commercial as commercial bed-nights in tourism in Denmark. Among Danish tourists non-commercial bed-nights are statistically dominant (Framke, 1995).

Among commercial overnight stays 35 per cent are in second homes, and this compares with camping (26 per cent), hotels (21 per cent) and holiday centres (10 per cent) (calculation from www.danskturisme.dk/web/analyse.urf/indkvartering). Second homes are privately owned, but many are accessible to tourists for weekly rental through various agencies and companies. What then are the characteristics of the networks involved in producing and using second homes (involving planning and legislation, private ownership and the business of renting them out)?

Owners stay in their second homes for holidays and weekends, and in some areas people commute to workplaces from their second homes during the summer. In Danish, the word for second home is *sommerhus* (literally 'summer house'). This term brings out the seasonality of second homes. To understand the meaning and production of 'homes' while on holiday (see chapter 1), we speak of 'second home' rather than 'holiday house', 'cottage' or 'chalet'. While second homes originally did not have heating, electricity and a water supply, most second homes for rent are now modern houses with such facilities.

Second homes in Denmark began in the late 19th century when they were built along the beaches north of Copenhagen by wealthy people (Tress, 1999, pp.91-109; 2002). The major boom in this construction was after World War II. Almost 175,000 houses were erected, 1950-1989. The number constructed peaked in 1972 with 9,000 built (Kaae, 1999, p.28). From the initial start in North Zealand, second homes spread to other parts of Denmark, especially coastal areas (Suadicani, 2002, p.193).

Today's number of 220,000 second homes implies that around ten per cent of households in Denmark own one. With an average of five beds (low estimate) per house, the accommodation capacity – in the period when one is legally permitted to stay there – is 300 million overnight stays. Only ten percent of this capacity is used for commercial or private use.

Danish legislation on second housing is strict in four ways. First, no private person may own more than three houses. Second, the construction of second homes for business purposes is not allowed; this is only allowed for holiday centres/villages. Third, foreigners are generally not allowed to buy such private property in Denmark. Fourth, during the winter season people are not allowed to stay more than 13 weeks. These four rules prevent a second home from becoming a legal business object, and secure decentralized property control. However, it is legal to broker second home rentals, and holiday house rental firms handle this. The accommodation capacity provided in this way is an important resource in Danish tourism.

Physical planning with rules for areas for second homes began in 1938 and developed during the 1950s and 1960s. The many plots, some as big as 2500 square metres, expanded the area used, and they can even be located in places regarded as beauty spots and worthy of conservation. Conservation laws and planning laws from the early 1970s introduced second-home areas as a separate type of zone from the rural and urban zones. This zoning produces a certain kind of tourist place favouring owners more than tourists renting the houses. Second-home areas do not have service centres or meeting-places, if there is no village or town nearby. Indeed, it is not obvious how these areas can be perceived as a 'destination'. For the tourists, it is the location of the house and plot that is somehow a 'destination'.

Furthermore, from a business or an economic-geography point of view, second homes have few 'destination' characteristics. Renting out the second home is only an additional source of income, and it is rarely the only or main reason for owning one. Most often, owners have little local commitment, and they are not members of local tourist associations. Owners of second homes are also tourists. Many houses are not rented out and many are lent to family and friends in informal ways, while some are rented out directly by the owner.

The major second home rental firms are national, and they monitor the quality of houses. The setting of the rent is the main variable available. There is little communication and face-to-face contact between owners and rental firms; little learning or any other kind of dynamic is involved in this. The position of second homes is very different from that of a 'destination'. Because of the diffuse ownership structure secured by law, business networks organized according to 'destination' cannot be detected. The dynamic networks are the weak connections *performed at a distance* between owners and rental firms, between rental firms and tourists, and between owners and their peers. In particular, the formal transactions involving rental firms are displaced; the only exception to this is going to the local tourist information office to pick up keys, or when any 'chance tourists' come to ask about a vacant house.

Second homes are highly material, localized facilities, with access governed by displaced networks such as virtual booking or connections with relatives. They are heavily 'policed' by legislation protecting a national resource and indeed perceived as a national symbol. Each house is a node of complex actor-networks of owners, family and friends, rental firms, tenants etc. Every one of these (up to 220,000!) nodes is an actant in producing tourist places in Denmark.

Although second homes seem to be a safe holiday environment in a period of terrorism, SARS and war, they are less attractive because of the downward size in nuclear families with small children, and the imperative of escaping the home that applies to other tourist types. Because of the decentralized ownership of second homes, tourist businesses, tourist organizations and local authorities can do little to enhance their attraction. Customers are not the only reason for the instability of this market. Renting out these houses depends totally on the varying interest and the situation of their owners. This form of accommodation is in decline (Tress, 2002, pp.119-20; Framke, 2001). Professionals in tourism are becoming very concerned since they are unable to revitalize this crucial market. They can produce brochures and websites where destination images frame offerings of houses for rent, but they can do little to secure demand or quality and maintenance of supply. The non-commercial tourist use of second homes is more secure than the business of renting them out. In this case, networks among tourists are more stable than those among tourist industries in producing tourist places, and it is not 'destinations' that organize such a tourism, materially and socially.

Deconstructing Destinations

Destinations name places that tourists regularly visit, but what more do destinations do in producing such tourist places? Destinations are first and foremost produced as images that are not territorially fixed. Rather than borders, it is symbols and identities that define them. This chapter has argued that destinations, apart from marketing, organize little of the networks that are so important in tourism.

Since tourism industries speak of tourist places as 'destinations', one can ask whether destinations are also imagined communities (Anderson, 1991). First of all, like the imagined communities of nations, destinations are characterized by little face-to-face contact between the actors communicating the images. Destinations are images produced by actors who do not need to meet each other directly, as the usual affiliations of marketing professionals, tourist organizations and tourists only concern a place. Secondly, destinations are communities that exist in the minds of the business actors and even the inhabitants of the place. To these actors, besides the obvious commercial purposes of place marketing, the destination represents a safety strategy that forms identities with a common destiny. As Bauman stresses, boundaries are not produced around existing identities: '...the opposite is the rule: the ostensibly shared "communal identities" are after-effects or by-products of forever unfinished (and all the more feverish and ferocious for that reason) boundary drawing' (Bauman, 2001, p.17). Boundaries, such as those imagined with 'destinations', are instruments in the formation of common identities to which local authorities and inhabitants may adhere. Destinations are constructed around historical representations and attempts at bordering; from the 'inside' destinations have an element of community-building.

Saarinen approaches a destination as a 'cultural landscape subject to continual transformation and reformation, in which it emerges, changes, disappears and re-emerges in varied forms' (1998, p.160). But this approach does not question the extent to which destinations actually work as the business units or the hegemonic discourses they are supposed to be. The identity of the tourist destination is approached as a '*discursive formation* which consists of what the destination is and represents at the time and the historical and present practices involved in transforming it' (Saarinen, 2001, p.51). Understanding the 'destination' as a socially constructed 'container' may suit the empirical context of concentrated 'one-place' tourist resort developments in Finnish Lapland, but it does not explain the complex actor-networks involved in holiday tourism along the coasts of Denmark.

The dynamic relations *among* firms in tourism cross the imaginary boundaries of destinations in multiple ways. If there are dynamic networks involved in the production of destinations, these are the crosscutting networks among firms and their employees, other local people, owners of second homes, and the visiting tourists. These networks are often informal, and not limited to the territory of the 'destination'. When they work at their best, they produce social capital that gives access to new resources (Bærenholdt, 2002). These resources tend to be about channels for establishing contact between service and experience suppliers and potential tourists. Such contacts are transmitted in relations different from those of the open market place. The network involving the repeat visitor, the café-owner, the second-home owner and so on produces personalized, cyclic networks of a community character that bear little resemblance to the official destination images. More than imagined identities, they are reciprocal relations among friends.

Compared with such relations, destinations are geographical images, which work at a distance, collapsing in proximity, where place-specific encounters matter. Word-of-mouth marketing at a distance in the personal networks of tourists transmits certain imagined destinations. It seems that the networks important to the *dynamics* of tourism are *not* those involved in the destination marketing organizations' efforts to pool their advertisements in destination brochures. But this does not render brochures or websites valueless; on the contrary, they are fundamental signs and platforms for communication in other networks.

Murphy stresses the benefits from the fragmentation of tourism. 'Fragmentation of the industry, which has been viewed as a barrier to comprehensive development and maximization of revenue, still retains some merit if it forms the basis for local diversity and character. The joy of travel includes noting regional variations in landscape and culture, as reflected in local architecture, customs, and food. Memorable visits can be made by the personal touch of individual owners and operators' (Murphy, 1985, p.36). Yet fragmentation does not necessarily produce attractive spaces as such. But if the attractive spaces of performing tourist places are those of diversity and crosscutting activities, fragmentation will be the consequence. Hence, this is also a dilemma for community-based tourism development that strives to organize destinations in accordance with tourist demand (see Murphy, 1985). Apart from rapid changes in demand that make 'making fit' an insecure business, tourists are attracted to the 'unfit', to surprises. This is especially the case with the flexible or mobile tourists who travel through multiple places and do not expect to get their whole experience within a single place.

Not only are tourist places like castles in the sand; the productive powers exercised through networks are as fluid as the sand used for the construction of sandcastles. It takes many networks to produce tourist places. If the 'adhesive water' of identification with 'destinations' fixes such sandcastles, they help to stabilize the place. In coping with the more or less unpredictable erosion, the networking of re-visitors and new visitors is the *sine qua non* of producing tourist places.

Chapter 3

Consuming Tourist Places

Experience derives from experimenting, trying, risking, the German *Erlebnis* and the Swedish *upplevelse* from living through, living up to, running through, being part of, accomplishing. Again the focus is on personal participation, we have to be both physically and mentally *there* (Löfgren, 1999, p.95).

Introduction

While chapter 2 investigated the uncertain role of tourist 'industries' and their networks in producing tourist places, this chapter examines how tourists produce places through 'consuming' them via sets of networked relations. Tourism comes to be stabilized through networks that need to be performed. Networking is a central aspect of tourists' consumption and performance of places.

However, tourist places are more than stages for tourist performances. The material qualities of places are also crucial. There are limits to how far tourist places can be fluid and performed to meet the expectations of all tourists. Tourists need to engage with the materiality of place. Tourists' consumption of places is a way of networking material, social *and* cultural elements.

To understand the specific role of place, the literature on shopping practices is illustrative. In Miller et al.'s study of two shopping centres, shopping is first characterized as a 'network of activity of which the actual point of purchase of a commodity is but a small part' (Miller et al., 1998, p.14). As in tourism, the networks include both 'hard' and 'soft' infrastructures. The former are the logistical systems for the production, transport, storage and display of goods, while the latter are the social relations among shoppers that govern their shopping practices. Furthermore, shopping is a routine learnt through socialization processes, but it is also reflexive.

While tourist practices and shopping practices both involve the use and production of consumer images, shopping while on holiday has qualities other than shopping at home. Tourist consumption involves the sense of performing 'routines' reflexively in *another place*. To shop in an 'ordinary' supermarket has a distinct flavour of the extraordinary for many tourists. Finally, shopping is about sociality (Miller et al., 1998, p.17), but shopping while holidaymaking is generally more about doing it in the company of others.

To sum up, compared with 'everyday' shopping, tourist shopping involves a stronger emphasis and reflection on the specific place as a space for *co-presence* and *memory*. Meanwhile, shopping is among the most popular activities on holiday, and this strongly influences the structuring of shopping space in such places (Snepenger et al., 2003). Even more than on everyday visits to shopping centres, tourists spend more time socializing than when just purchasing commodities at home. A study of urban tourism in Benidorm showed that tourists stroll around the town for an average of three and a quarter hours a day. Strolling is a tourist attraction itself in such urban 'vacationscapes', and 'the best theme parks are towns' (Iribas, 2000, pp.112-13).

Most tourists do not take the place for granted as they do their home town. For example, when staying in second homes, tourists consume places that are hybrids of home and away. Tourists attribute meaning to different dimensions of place, but the performances of tourism are always about consuming *places*.

The next section discusses five principal dimensions of tourist places which are used to organize the subsequent case study of strolling tourists consuming a rather ordinary place, a maritime town on Bornholm. The chapter investigates how material natures, social relations and cultural conceptions intersect in the performance and stabilization of tourist places in time and space.

Dimensions of Tourist Places

'Place' is often associated with the home community. But in the modern world people attribute meaning and relate to many places – their place of birth, the place where their parents' home is, good friends' places, sacred places, places central to national history, or holiday places. People try to make sense of these by performing them in accordance with meanings attributed to them. Yet people attribute different meanings to the 'same' place.

Tourist places are simultaneously places of the *physical environment, embodiment, sociality, memory*, and *image*. Tourism depends on these diverse but also overlapping notions of place.

The idea of *place as physical environment* corresponds to place as material. In traditional tourism research, place has been seen as a matter of the spatial position of the 'destination' (see chapter 2), and in geography places are identified as dots on maps, places as locations (Agnew, 1987). But a better way is to follow Ingold's suggestions that place is always an environment for somebody, that, it is continually under construction, and that environment is experienced from *within* and not something that humans are outside (Ingold, 2000, p.20). Places are historically produced material environments. Rather than the territorial character of locations, it is the human, bodily engagement with the material world that is important to place as environment.

The notion of *embodied place* has emerged in recent research. Bodies encounter places and are 'surrounded by place' (Crouch, 2002). 'We live places not only culturally, but bodily' (Crouch et al., 2001, p.259). The corporeal approach to place makes it possible to understand places as practiced, produced, and performed. The movement of tourist bodies can be seen as prescribed, learnt and regulated, but 'bodies are not only written upon but also write their own meanings and feelings upon space in a process of continual remaking' (Edensor, 2000b, p.100). Embodied place includes other temporalities than that of the abstracted natural histories/physical geographies of the environmental. Together with the time-space of flows of water and matter, are the rhythms of bodily mobility and proximity. Thus places are produced not only in space but also in time. This approach needs to 'temporalize place' (Crang, 2001, p.204), and this is stressed by people's experiences of places important to them. For example, different temporalities and spatialities characterize various momentous moments in people's lives.

Conceptualizing place as embodied also enables us to understand *place as sociality*, and vice versa. The 'togetherness' or co-presence of people is central to the definition of place as 'locale' (Giddens, 1984; Agnew, 1987). Sociality and materiality therefore overlap in the production of contexts and contents of social practice. This is more than a question of the co-location of bodies. Social interaction is materially organized, whether the proximities involved are 'bodily', 'virtual' or 'imaginative'. 'Virtual proximity' is performed in the mobile networks of cyberspace, while 'imaginative proximity' depends on the technologies that bring faraway images into people's living rooms (Urry, 2000). However, performing bodily proximity is central to the social practice of consuming tourist places. The performance of face-to-face (and body to body) sociality contributes to producing particular place for a period. Central here are the social practices of meeting, chatting, feeling and seeing one other. The sociality of place ranges from being part of the crowd to the intimate interaction of love (see chapter 8).

Place as sociality is multi-dimensional. It involves mobility and connections with other places. Sociality is embedded in micro-scale movement, and various means of transport are present as constant reminders of how one arrived. Besides the social relations inside the family/group/couple, few of the relations are stable. In her work on rural artists' colonies around 1900, Lübbren points to the importance of the sociality produced among the group of artists who formed specific colonies with their 'anti-tourist' practices. The sociality of their gatherings and networks produced a way of securing the freedom to perform their specific mode of living (Lübbren, 2001, p.19). Thus the social networking of the travelling group is both a goal of sociality and a means of securing freedom. It involves certain internal orders in time- and space-regulating practices, so that people can meet one other. It stabilizes certain forms of social ordering.

The meaning of tourist places also depends on tourists' remembering their experiences. *Place as memory* produce places in the double temporalities of memorized time and the time of memory. Tourists categorize, compare and consider different qualities of places experienced at different times. 'We think over where we have come from and how far we have come and where else we desire to go. We negotiate an awkward slope. We make little judgments, reflexively we talk things over, mixing and re-mixing all of these impulses and desires. There are particular things in this place, another encounter, recent memories of similar places and what we did there' (Crouch, 2002, p.211). The production of memory is grounded in the reflexivity of increasingly mobile tourists who have experienced many such places (Urry, 2002). As already mentioned, we need to understand places as temporal.

The most crucial aspects of place as memory are 'small worlds' of care. In humanist geography:

...places are not only highly visible public symbols but also the fields of care in which time is of the essence, since time is needed to accumulate experience and build up care. All places are small worlds: the sense of a world, however, may be called forth by art (the jar placed on the hill) as much as by the intangible net of human relations. Places may be public symbols or fields of care, but the power of the symbols to create place depends ultimately on the human emotions that vibrate in a field of care. Disneyland, to take one example, draws on the capital of sentiments that has accumulated in inconspicuous small worlds elsewhere and in other times (Tuan, 1996, p.455).

Tuan's 'fields of care' give a sense of the sentiments and emotions invested in place as memory and the social relations associated with it. Places matter to people because they are meaningful 'When tourists experience places, their experience is not so much directed by what they encounter, but by what meanings they give whatever they do encounter. That in turn, derives from their intentions towards the places, the framework with which they understand them, their situation as visitors, and the extent and nature of their involvement' (Suvantola, 2002, p.33).

However, the humanist geographical approach gives little understanding of how emotions and identifications relate to embodied and socialites, because it assigns priority to relatively fixed meanings. This approach should be complemented with an understanding of how memories are socially produced *with* other people. Memories are the product of cooperative work and are embodied (Urry, 2000, p.136; and see chapter 6). 'Place as memory' is performed in social and embodied relations between people with associated mobilities and proximities. The memory of a place can be tied to the effort of actually getting there: 'we are inclined to recall the road we took' (Ingold, 2000, p.204). A place is also often remembered if one meets somebody there that you either already know or come to know in this way. It becomes the place of the coincidental meeting, and it does not need to be a meeting that is necessarily enjoyed. In other words, memories can be bound to specific doings-in-place: the souvenir purchased, the obligatory accident walking with the ice cream, the more or less desirable meeting with somebody, the meal consumed, the bus missed, the necessary bicycle repair and so on. These are all doings that structure the temporality of experience-in-place. Furthermore, arranging the family album in the chronological order of events before and after the visit may also structure these memories.

Place as memory is bound to the performances and experiences of place as sociality and embodied place; the opposite is also true. The bodily and social performance of place draws on earlier memories of earlier visits to this or other places, while memories are contested and revised in hermeneutic circles. Tourists also produce materialized images, and consume the images supplied to them in complex intersections of circles and spirals of reinterpretation and performance of places.

Place image is the final aspect of tourist places. Recent cultural geographies and cultural studies have emphasized the production of place images (Shields, 1991; Jackson, 1989; Gregory, 1994; Crang, 1998). Shields defines place images as 'the various discrete meanings associated with real places or regions regardless of their character in reality. Images, being partial and often either exaggerated or understated, may be accurate or inaccurate. They result from stereotyping' (1991, p.60). In this way, images work as more or less free-floating signifiers, independent of the signified.

It is crucial to make a distinction between place images and place as memory. Tourists' sentiments, dreams and imaginations have other qualities than stereotyped and more or less dominant 'representations of space'. To follow Lefebvre (1991), place of memory as well as social place has to do with 'representational space', whereas environmental and embodied place are 'spatial practices'. There are dialectical relations, between the three spatial dimensions in Lefebvre's work and among the five dimensions of place suggested here. For example, narratives are more than just memories, since travel memories are also performed in travel stories. Narratives are performed, making sense of practices while performed, while memories are reactivated and reflexively reconstructed. Memory is an integrated dimension of the performance of tourist places. Expectations of the future, memory of the past and attention to the present are interconnected (Simonsen, forthcoming).

We now exemplify how the environment, embodiment, sociality, memory and image are performed in the consumption of one particular place.

Strolling around a Seaside Town

In the well-developed tourism on the island of Bornholm, the attractions include nature/landscape, beach life, seaside walks and museums. But visitor surveys show that a further attraction is that of 'atmosphere' (Nilsson, 2002). A row of four seaside towns on the rocky northern coast of the island is visited by many tourists because of their distinct 'atmosphere'. These small towns around a harbour were once dominated by fishing. We focus on the 'atmosphere' in such town Allinge in northern Bornholm. It has a few thousand inhabitants and the harbour environment is comparable to other sites on Bornholm. Yet the other sites tend either to be oriented solely towards tourism (Gudhjem and Sandvig) or are more famous for the preservation of their historic buildings (Svaneke). Allinge is least marketed as a tourist destination, and has most permanent residents.

Tourists stroll around all these towns, but Allinge is particularly good for strolling or flâneurie. There is a concentration of souvenir shops, harbour, smokehouses and supermarkets within a very small area. Allinge is also a quite ordinary place because of its role as a service centre with many non-tourist shops and transport connections. It is a place for shopping in order to consume the place. The presence of supermarkets similar to those tourists know from home is central to tourists' stays in second homes and camping sites. Allinge is a good location for studying the ordinary extra-ordinariness of tourist consumption.

Ongoing tourist performances were observed for two days in summer to analyse patterns of movement, what tourists do and how. Tourist performances were photographed. On another day 16 interviews were conducted among the strolling families/couples from various countries (with an interview guide similar to the one used for the study of tourists on the beach, presented in chapter 4). There is also an participant observation. Interviews were also conducted with a municipal planner and with the harbour master and his assistant. Tourist brochures, tourist/historical literature and planning documents were collected. The analysis is structured along with the five dimensions of tourist places presented above.

Place as Physical Environment

Historical geographies are helpful in investigating the material character of place as environment. Bornholm has a distinctive geological history with the production of non-horizontal layers of granite and sandstone intersecting with varying sea levels. Granite dominates northern Bornholm, and this means that the island presents an environment of rocky coasts where natural harbours have been one of the determining factors for the development of settlements, fisheries and so on.

Allinge, the 'capital' of northern Bornholm, has one of the major natural harbours on Bornholm, but it owes its specific tourist qualities to construction work. After seven years of construction the inner and the outer harbour were finished in 1862, then destroyed by floods in 1872, and expanded in 1884 and 1914-19. The development of granite quarries by German businessmen paved the way for tourism (Rosenquist, 1991), just as their networks spread knowledge of the place as a tourist attraction. Ships and ferries with direct connections to Sweden (historically also to Denmark) have used the outer harbour since 1895, although there have been many changes in the routes. On shore, Allinge had several merchant houses and farms in the town. In some of the houses along 'Havnegade' (Harbour Street – see Figures 3.1, 3.2, and 3.3), a number of seasonal shops and restaurants were opened as a result of tourism (Rosenquist 1991).

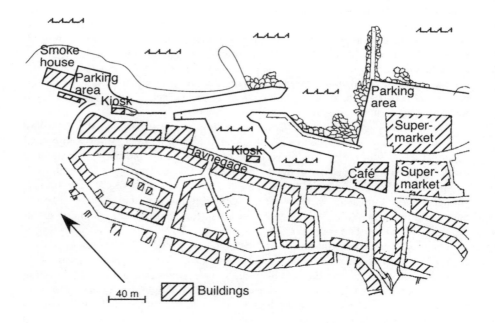

Figure 3.1 Map of Allinge Harbour

Figure 3.2 Harbour Street, Allinge

Figure 3.3 Allinge Harbour towards Harbour Street

In the 1970s, the remaining equipment from the granite industry was removed, and during the 1990s fishing, which used to dominate the harbour, disappeared with the crisis in the Baltic fisheries. Nowadays, yachts from many countries occupy the harbour during the summer. The excursion boat for Christiansø, Denmark's easternmost island, leaves once a day in summer, and from time to time catamaran connections to Poland are active.

There are also kiosks serving ice cream, sausages and burgers, the restaurant 'Algarve' (Figure 3.3), a pizza bar, and a French-style cafe (Figure 3.5) with a fine view. A smokehouse (Figure 3.4), a small pottery workshop and parking areas are nearby, in addition to closed fish factories. It is a physically compressed space. Harbour Street is a bottleneck with passing cars, bikes and coaches, and projects have been proposed (and implemented in 2003) to make the street 'a rustic pearl' instead of 'an asphalt dessert' (*Havnegade, Et bymiljø projekt i Allinge*, p.2).

Tourists consume Allinge harbour multisensuously by walking around and shopping. If they buy anything it is smoked fish, ice cream and other refreshments, souvenirs, clothing, and possibly food from one of the supermarkets. The weather is crucial and is often discussed. Harbour conditions are critical for sailors when the wind is in certain directions. Waves are often seen breaking over the piers. In good weather, tourists can view most of the Baltic Sea traffic (supertankers, cruise ships, naval ships etc.). Weather is often discussed by tourists in their outspoken reflections on whether to go to the beach, the museums, the nearby fortress of Hammershus (see chapter 5), coastal paths or for a stroll at Allinge Harbour, while sailors consider when to cross the open sea to the Swedish coast.

The tourists' consumption of Allinge's place as environment engages with complex flows on various temporal and spatial scales in various contexts. These are the geo-historical complexities of the emergence of granite, seas and weather; the networks, flows and the remaining built-up environments of granite quarries, fisheries, agriculture, trade and transport; the mobilities and proximities of tourism development; and we must not forget the practices of the inhabitants. These historically-produced material characteristics represent the stability and the 'physical evidence' of the place as environment.

Embodied Place

Place as environment is also embodied, and vice versa. The coexistence of multiple corporeal performances produces tourist places. Bodies are part of the environment, and the environment is embodied.

Strolling practices are varied. Many middle-aged couples, so-called 'empty nesters', visit Bornholm and Allinge Harbour because of the atmosphere and spend much time here. Families with children and many others choose to stay at the beach all day on the sunny and less windy days, visiting the town of Allinge on less sunny days. People discuss how to arrange their days on Bornholm in order to get the best out of the holiday.

A Danish couple in their fifties stand on the pier and gaze at the boat departing for Christiansø. They are 'looking around at the sea, the town, the harbour'. They say they

Figure 3.4 Smokehouse, Allinge

Figure 3.5 Allinge Jazz Festival at the Café

would like to go to Christiansø on a day with better visibility. Today they 'browse around and look for an open sandwich ... you know, a beer in a cafe, drifting around, isn't that what it's all about?' Such tourists not only drift around on foot; they also drive casually around the island visiting several places each day. Another Danish couple in their fifties say 'we potter about' and a Swedish couple in their twenties say they are 'taking it easy'. People walk with a relaxed attitude, occasionally visiting shops, and the consumption of food and drink is one of their organizing modes. The type of walking they perform is that of 'strolling'.

The micro-mobilities of strolling and consuming the harbour and the shops in Harbour Street starts around 10 a.m. and continues until around 6 p.m., but on some days, especially during the jazz festival, there is plenty of life in the evening. A family usually spends around an hour on the stroll, including shop visits and the purchase of ice cream and drinks in the kiosks. The total number of strolling tourists from the cafe in the south to the smokehouse in the north varies (from day to day, with the time of the day and the weather) from 100 to 500 at any one time, while the number of people in 'front-stage' jobs is around 35.

Much strolling is a spin-off from other activities. First, the supermarkets are the instrumental reason why many people come to Allinge. When the car or bicycle has been parked, people can take a stroll before or after shopping. Many visit Allinge just to eat the famous 'Bornholmer' smoked herrings for lunch at the smokehouse (see Figure 3.4). Tourists stroll to prepare for or finish off food consumption, often with an ice cream, and the arrivals and departures of ships on the Sweden, Christiansø and Poland routes also produce tourist movement, among the travellers themselves and among those gazing at the event.

Other occasions for a stroll may be events such as the annual jazz festival in July at the harbour (Figure 3.5). The festival is arranged by local associations and sponsored by local businesses. Many tourists and locals meet year after year on this occasion (this was observed several times by the researcher). The music from outdoor stages adds to the atmosphere of the whole harbour area, especially because of the acoustics of the compressed and rather close space between the buildings. It is mainly middle-aged and older people who 'inhabit' the benches, tables and beer booths in front of the stage, while others look down on the event from the raised platform of the café. The temporalities of the music programme structure the flows of people. Breaks produce dispersal, while the introduction of a new band draws them back again.

Other events include the arrival of between 2000 and 2500 yachts a year. Since space is limited, yachts have to be efficiently organized and placed. Following an arrival or departure, yachts must be reberthed. The harbourmaster enacts a show where he plays the main role himself, shouting his commands in the Bornholm dialect, and yachts are rearranged with their crews jumping around. Not all crews are so highly trained in harbour navigation, so commands and gestures are crucial. A crowd of people watch, having a chat with the harbourmaster in between.

Much of the life on board the yachts is easy to see from the piers and raised streets around the harbour. A characteristic gender division is noticeable along Harbour Street. While women browse in the shops, men gaze at the yachts and life aboard them. Meanwhile cars, bikes and buses trundle through Harbour Street at slow speeds.

A male Danish (multimedia) student in his twenties explains 'I saw the place from the bus and got off', interpreting the qualities of the place as 'relatively authentic'. He regards it as 'fantastic' and emphasizes that the signs are not only in foreign languages. Hence the 'authenticity' of this rather ordinary place is found in the signs of the presence of local people and non-foreign tourists. As it is a rather ordinary place, and certainly not a place for sightseeing; photography is almost absent apart from a few, semi-professional photographers. Yet because video is better for recording sound and atmosphere, video filming is performed at the jazz festival.

Tourists 'botanize' the shops, refreshment facilities and harbour environment producing a tourist place. Few would disagree in calling this 'one place' because of the compressed space where interactive communication with speech/shouting, music, appraisal, gesture and eye contact is possible around and across the harbour.

Place as Sociality

Proximity is a condition of sociality, and bodily proximity is a condition of the type of sociality which has more durable effects on minds and memories. Some places as environment facilitate bodily proximity more than others. In the compressed and condensed space of Allinge Harbour, where flows criss-cross, people cannot escape experiencing congestion. There are many forms and orders of place as sociality; or, more precisely, there are several social places around Allinge. One social place is that of tourists dining in their yachts, another is that of tourists in a passing coach glancing at the tourists dining on them. The material order of the environment is perceived with specific qualities. Specific meanings are ascribed to it and this means that social places overlap and also share meanings.

A family from southern Sweden has crossed the Baltic in a couple of yachts, something done ten to twenty times before. The parents, in their forties, state that 'Allinge is special, the feeling ... we feel at home, have direct access to the smoke house and shops, proximity to shops, a Whisky Steak at the inn, fine supplies and the busy harbourmaster'. That Allinge is special explains why they return again and again.

First, then there is *the atmosphere* that is produced by the complexities of tourists, locals, traffic, shopping and so on (see chapter 8 on atmosphere or hauntings of a place). Social practices produce sociality *in general*. An extended German family stays for periods varying from two to four weeks in a second home in southern Bornholm. It is the sixteenth visit of the parents, and they praise the landscape, the silence, the pine trees and the sand on the beach at Dueodde where they stay. They also enjoy Danish people and the Scandinavian countries in general. The place is 'not too full, not so crowded, it has a relaxed atmosphere'. It is a tradition for them to visit Allinge every year. The mother 'goes to the 'Warehouse' every year'. At the time of the interview, they have just visited Hammershus (see chapter 5) and are going to visit other places with glassworks and shops, in addition to the obligatory shopping for smoked salmon and herring. For this and many other families, the general atmosphere of Allinge is synonymous with the atmosphere of Bornholm in general; the site is part of the place or destination. The performance of place is bound to the insularity of the destination, where the social life is also something generated among the tourists who perceive it, the tourists are co-producers of the atmosphere of sociality.

Second, there is the sociality of *specific events or practices*, such as being in the audience at the jazz festival or consuming the fish buffet in the smokehouse (see Figure 3.4 and 3.5). Compared with the complexity of coexisting practices producing the general atmosphere, these are more focused, instrumental and simple from the outset. But people listen to music, which also includes gazing and feeling, and eat lunch in specific environments of embodiment and sociality. So the second-order place of sociality is about certain *doings*. But they are performed in coexistence and proximity with other and diverse forms of doings, all adding to the general atmosphere. It means a lot for the atmosphere of Allinge whether or not the outdoor tables of the cafes, the restaurants and the smokehouses are 'inhabited' or not, just as the pleasure of guests in these facilities may well depend on the other performances going on around the harbour.

The third order of place as sociality is made up of the *specific encounters* of people meeting one other, the yearly 'hello again', chatting with the harbourmaster who always seems in a good mood, and the criss-crossing encounters resulting from his role as mediator. Encounters with other tourists and especially with locals are considered important, but not all tourists dare to do this. Encountering the 'authentic' locals is often a much appreciated tourist experience. The Tunisian male member of a visiting Dutch-Tunisian couple in their fifties stressed his experience of meeting and talking to a fisherman in another small harbour. The Dutch woman had her first vacation on Bornholm when she was only five years old, so today she knows the island well. They have come to Allinge Harbour to shop and especially to order fish in the smokehouse. But they miss an active fishing harbour and especially shops offering local fresh fish that has just been landed (as they find in Tunisia).

The social interaction within the *travelling family or group* is the fourth order of sociality. This is clearly very central to many families with children. The three genera-tions of an 'extended' family visit Bornholm at the same time so that they meet one another at a variety of places. While during the day they may experience different things according to their interests, they meet for some meals. An interviewed couple in their fifties form the older generation of grandparents and parents in this family network. Besides art, they like studying the menus of restaurants providing local food, investigating where to organize meals they can share with their children and grandchildren.

The four orders of sociality intersect, and for some visitors the social atmosphere or specific events do not help the sociality within their own group. This is the case with one young Danish couple, staying at the house of the man's father. His girlfriend experiences the place as 'rather cosy' since there is a shortage of younger people. The shops and jazz festival appeal more to the older than to the younger generations of visitors.

Places of sociality of orders ranging from the general atmosphere to the family/ group/couple intersect in numerous ways. In some couples/families, people are together but do different things. The sociality of the general atmosphere is produced by the complexities of diverse tourist and non-tourist practices, and the social performances of other tourists are central. Shopping and strolling are performed in places with 'enough' but not too much of a crowd. The atmosphere of the harbour is perceived and performed in relations with the image of the destination. The networks governing flows and mobilities are varied, and they work within the specific cultural languages of social groups, age groups and so on. Teenagers' sociality works in networks different from those of 'empty nesters'.

The place of sociality is vital to the performance of tourist places. To a great extent, tourists themselves produce this distinct atmosphere. Shopkeepers, service personnel and guard-figures like the harbourmaster are important 'front-stage' performers, but the meaning of their practices depends on the co-present performances of tourists. A place, a restaurant or a shop is not attractive consumption if it does not meet tourists' expectations that it should be crowded, unless a romantic desire to dine alone is stronger than the need for collective sociality. Likewise, it is tautological to say that *events* that do not attract an audience do not signify a well-performed tourist place, and *encounters* also depend on the active participation and often initiating practice of tourists. Tourist places gain their status through those performing the place with the qualities they attribute to it. This is the economy of tourist places, an economy of performance.

Place as Memory

The performance of tourist places is bound to the intersecting temporalities of how people remember former visits, how they perceive the present, and what they remember and forget in the future. The sense of place involved in the selective processes of memorizing comes from multiple senses, including the experiences of bodily movement and of sociality in place. Photographs may facilitate specific things to be remembered; the selection of what to record is a selection of what to remember in future.

Even though few people perform photography in Allinge Harbour, memory is at play in several ways. First, *historical photographs* are staged. The Harbour Office is full of historical photos of visiting ships, of flooding and the rebuilding of the harbour. Few tourists come into this office, which belongs to another, past world, but they may see photocopies of the same scenes decorating the walls inside the cafe. The cafe (called 'Højer's Ice Cafe') also has historical photos of the Hotel Højer that used to be in the building where the cafe is now, so that tourist history also becomes part of presented memory in-place.

The place is also part of the memories of *revisiting* tourists. A Danish family who visit Bornholm every second year enjoy the diversity of the island and 'always see the same things'. To the children, to stroll around Allinge Harbour is to have a nice time looking for souvenirs. Revisiting tourists are common on Bornholm, and they perform with certain embodied mindsets of how this place is and how it should be visited. To revisit the place is also to assure oneself of the *durée* of the place, and if there should be any change, however disturbing, it can be accounted for. Sentiments from elsewhere are brought along; they form principles for doing things the family normally do on holiday. In addition, memories of place are also memories of growing up and education: do you remember how you did that at that age/in that year? Places are assigned meanings from the small worlds of people's biographies.

Yet another type of regular visitor is represented by a Swedish couple in their fifties. They have come to Bornholm by boat every year since the late 1960s. This year they have taken another couple with them to show them Bornholm. Showing others 'your' tourist place is the way tourists really take possession of place, to the satisfaction of both the guides and the 'guided'. They eagerly accounted for the attractions of Bornholm: 'the smoked herring, the quiet, the sea, small fine harbours, the "picturesque"'. There were very specific shops, smokehouses and places they had to show the others, and they stressed their knowledge of and direct personal relations with people who ran specific tourist attractions. The attraction of 'Arne's' smokehouse and fish shop in Allinge was also much associated with the owner Arne, who was present in the shop. Encounters are crucial to the production of memories, so people looked for them again and again.

Memories are associated with items bought more than with photographs. Fish food is consumed in-place and bought in packaging suitable for transport home or even mailed to family and friends. It does not seem to matter that the herring smoked on Bornholm is imported lorries from other parts of Denmark. The memory of place is also about the *taste of food*, and such tastes are mobile.

People remember some experiences and performances and forget others. In addition, there is the 'biased' or 'coloured' memory. For example, while writing this chapter the researcher remembered the houses in Harbour Street as half-timbered, because this is the image of the typical picturesque Bornholm town house. However, photographs (see Figure 3.3) showed that this was not the case. Memory is massively influenced by photographic images.

Place Image and Beyond

Places are produced as images in texts and pictures: in tourist brochures, guidebooks and postcards. They transmit what the so-called local tourist industries think might attract tourists. So how do they do this in the case of the rather ordinary place of Allinge Harbour?

The commercial tourist brochure *Turist-Bornholm Guide 2001*, financed by the advertisements in it, says nothing about the place, yet many Allinge shops advertise in it. 'Town walking' in 'the old market town's narrow streets with exciting old houses' is mentioned (p.84), but with reference to other towns. In the pages on Allinge and neighbouring towns it is mentioned that 'Bornholm's tourism started in northern Bornholm...' (p.60), but detailed descriptions are about 'natural' environments. In the tourist guidebook *Turen går til Bornholm* (by Søren Lauridsen and Kaj Halberg, Politiken, 1997) Allinge's history and facilities ('the shops, kiosk, library, post office') are accounted for in a rather enumerative descriptive style. The local *Turistinformation for Allinge-Gudhjem*, published by the Allinge-Gudhjem municipal association of tourist and industrial firms, also gives this kind of presentation of facilities, describing the shops as 'very well-stocked' and accompanying the description with photographic views of the harbour area. These photos are always oriented towards the characteristic buildings – the 'Warehouse' or the former 'Hotel Høier'. The Bornholm transport company's brochure *Around Bornholm 2001* directs attention to other parts of Allinge with 'small streets' and 'old farms and historical houses' and does not mention the harbour area and its shopping facilities.

The attraction of this very specific place is not much accompanied by circulating images. The site is not a sight. But people remember and revisit the compressed space around a harbour with its narrow passage to the sea where ships come and go, and its light and atmosphere constantly change with the weather. In memory too the buses, cars, bikes and pedestrians drift through Harbour Street, where speeding traffic is not appropriate either for passengers 'glancing' out or for pedestrians and cyclists. The shopping stroll, the consumption of the Bornholmer herring and the contemplative hanging around in the cafe are further ways to perform this place. The relevant place image portrays the more general Bornholm atmosphere that destination marketing material often transmits. But the most significant personal memories of a place can be other than those of place images, for example from childhood (Edensor, 1998, p.146). Many performances, that produce places, lie beneath the surface of images, rather like an iceberg or as ghostly hauntings (see Degen and Hetherington, 2001).

Strolling and Stabilizing Place

How then to understand the modality of strolling, drifting, hanging around, shopping, eating, and chatting around a 'non-sight' site? What senses are involved in consuming such a place? How are these practices stabilized?

The complexities of the performances that produce such a tourist place are considerable. Allinge Harbour is not a non-site because its quality is its inviting sociality. It does not look like a stable place; most people are on their way to the beach, to the supermarket, to the smokehouse, to the excursion boat and so on. In spite of this gateway character, the place is not a 'dead time'. It is a place for having a good time, taking it easy, and with Bachelard we can note that 'it is only in fluid (liquid) place that conflicts are dissolved or transcended' (quoted Buttimer, 2000, p.212). Tourist movement is stabilized by the materiality of place as environment, and on the other hand by the networking involved in the embodiment, sociality and memory of tourist places. We show how this happens by reviewing the five dimensions of place.

The materiality of *place as environment* involves the trace of complex temporalities and spatialities, where many features are solid and enduring relative to the time span of human life. Among the enduring properties are the land areas, buildings, roads and harbours, many of which need maintenance from time to time to be performed as a tourist place. Place as environment involves the mutability of the weather and the sea. The temporalities and spatialities of these crucial environmental dimensions are embedded in global flows of matter usually approached as circulation systems with features of the chaotic or the unpredictably complex. Still, the materially localized character of physical environments stabilizes places, thus ensuring that places actually can be revisited. Furthermore, the relatively enduring character of environmental place is often presupposed as 'a fact' by tourists' imaginations and memories, as well as in place images. In fact, tourist places change, but tourists celebrate that this only happens slowly.

The performance of *embodied place* cannot be separated from the place as environment. The sea itself is a limit; a limit often challenged by tourists, not all of whom survive. Likewise, traffic is an opportunity, but also a danger to be coped with. Roads, piers, tracks and buildings are sediments of embodied practices over generations, and these are practices of people from here and there, networked in hybrids of territorial and mobile spatialities. The specific strolling routes intersect the environment in constant remakes of the performances in response to weather conditions, but also to the times of day when accessible facilities such as restaurants are open, in addition to the temporalities of the chance of finding a parking space for the car. Bodies within the network or group of performing tourists emerge in constant readings of conditions, and are thus also produced by the actions of the very same network or group. Tourist consumption of place is stabilized by people's regular demand for food and refreshments, so that the supply facilities become nodes in tourist mobility. Furthermore, restaurants, kiosks, smokehouses, and supermarkets are not coincidental nodes, they are networked by memories of consumption performed earlier by people or their peers.

Sociality of different orders – from the tourist couple, family, group or network to the general atmospheres of places – often facilitates moments of pleasure and fun; but these moments are more than social. Remarkable tourist moments are hybrids of environment, embodiment, sociality, memory and images. But they would hardly be triggered off if they did not include the social relations of attention to one another, the feeling of one another's presence and the resonance among individual performances. The presence and attention of other people make strolling meaningful, for better or worse. The sociality of landscapes is about how we '*also attend to one another*' (Ingold, 2000, p.196, his italics).

The notion of attention stresses the sociality of the coordination of actions to a higher degree than simple presence. This means that strolling is not only performed with a consciousness of the presence of others, but is a performance in interaction. 'By watching, listening, perhaps even touching, we continually feel each other's presence in the social environment, at every moment adjusting our movements in response to this ongoing perceptual monitoring' (Ingold, 2000, p.196). To illustrate the role of social interaction, Ingold uses the metaphor of the orchestral performance that builds up and resolves tensions, networks multiple rhythms, and only makes music by performing it. Like music, tourism only exists through performances. Strolling can be recorded in films, but this hardly replaces the experience of bodily strolling, because strolling is an experience of interactivity with the present environment and the attention of other people in constant adjustment and 'spatial negotiation'.

Tourists' attention to one another is a networking practice with effects beyond the moments of tourist consumption themselves. Networking extends in time and ensures revisiting and re-meetings of tourists. Meanwhile, the complexities of images, materialities, businesses, tourists and locals may become destabilized, so that networks fade away and tourist places change or even become 'dead' places of the past. This is so because tourist networks are highly ambivalent. They are attracted by places, but they may also quickly change their routes. The consumption of tourist places involves the combination of involvement and detachment characteristic of fluid cosmopolitan life (Urry, 2003, p.137). Also tourists are involved in proximate and home-like relations, where the sociality of attending to one another endures in time among familiar actors in forms of territorial spatial practices. Tourists detach themselves from the bonds of such experiences and perform other places.

Place *memories* are bound to specific events or encounters performed. The memory might be of the atmosphere of drifting around on the very warm evening when 'we' met another family and music was played. The 'we' indicates that the memory relates directly to the specific sociality of the group – the family, the couple or the like. It is a memory easily recalled when one revisits the same place on a cold day, or revisits other places on warm evenings with music, or meets the same family in another place, or when one looks at the photographs of activities performed earlier in the day on the way towards the warm evening, although no photographs were taken that evening. Memories have to do with sequences of actions in time and space, and may include the memory of place associated with the memory of a person who has died since the meeting. Place as memory is tied to multiple pasts, presents and perhaps even anticipated environmental, embodied and social places 'here' and 'there'. Memories network places over time.

Various place *images* and myths may be involved in the enactment of place, but the most powerful myths are those not scripted into places but incorporated within people's memories of tourist practices in the same and related places. To consume tourist places in certain modalities means to practise embodied haunting dispositions. Meanwhile, performing these practices is part of complex networking activities that compose the interfacing rhythms of various orders of sociality across space and time. Tourist places performed this way have the same attraction as music. There can be recordings and memories, but places only play the music of tourism as long as they are performed as such. The sociality and rhythm of attending to other people is crucial to the consumption of tourist places. If the music performed in sociality becomes memories, it may also stabilize tourist places over time.

Chapter 4

Staging the Beach

The edge of the sea is a strange and wonderful place. All through the history of Earth it has been an area of unrest where waves have broken heavily against the land, where the tides have pressed forward over the continents, receded, and then returned. For no two successive days is the shoreline precisely the same. Not only do the tides advance and retreat in their eternal rhythms, but the level of the sea itself is never at rest. It rises and falls as the glaciers melt or grow, as the floor of the deep ocean basins shifts under the its increasing load of sediments, or as the earth's crust along the along the continental margins warps up or down in adjustment to strain and tension. Today a little more land may belong to the sea, tomorrow a little less. Always the edge of the sea remains an elusive and indefinable boundary (Carson, 2000, p.2).

Introduction

We have discussed how tourist places become produced, stabilised and consumed, and how tourists' performances are central to those processes. In this and subsequent chapters we examine in more detail the practices that tourists deploy in performing tourism places. Our point of departure will be a site that more than any other is tantamount to the tourist life of leisure, the beach.

The beach is a classical trope. To the affluent classes of the post-war period in Western Europe the 'global beach' is the very hallmark of leisure and tourism (Löfgren, 1999). The perils and attractions of beach life in the tropical and sub-tropical climatic zones of the planet is a recurrent theme of late 20th century popular culture. Throughout this period the holidayscapes of the Mediterranean and the Caribbean has been among the foremost magnets of global tourist flows (see Sheller, 2003). These flows were generated by and contributed to, a cluster of fantasies and desires revolving around the exotic otherness of paradise beaches with a more or less explicit erotic character (see Herold, Garcia and DeMoya, 2001).

Historically, the beaches of Northern Europe carry a less hedonistic but no less significant connotations. Of particular importance has been the construction of the seashore as a source of health and tranquility. The temperate beaches of the North were culturally formed in a binary relation to the supposedly unsound and morally corrupting 'sultry climates' and 'lukewarm waters' of the subtropical Mediterranean (Littlewood, 2001; Corbin, 1994, p.104ff). Hence, the fresh air and the ever changing yet unchanged seascapes played a crucial role as an antidote to urban everyday life throughout Northern Europe. This construction of the beach as a symbolic 'other' to both urban life and the hedonistic beaches of the South were instrumental to developing tourism in Denmark.

This chapter discusses the cultural and material production of the beach as a stage for tourism performances. We examine how 'the beach' is (re)produced through a multiplicity of social and material practices as well as the enactment of contradictory cultural discourses. In its divergent forms beach life is a thoroughly scripted, staged and performed practice. It is played out on a stage including the backcloth of the sea, the radiant sun, the textures of the sand, as well as set pieces and accessories such as fishing vessels, mobile phones and cars.

While the metaphor of the theatre thus seems an appropriate way of grasping tourism performance it also carries some problematic connotations regarding space. 'Too often', Coleman and Crang points out, 'dramaturgical metaphors suggest performance occurs in a place – reduced to a fixed, if ambient container' (2002a, p.10). Hence, the multiple ways in which the 'stage' for tourist performance is *produced* is left out of focus (see chapter 1). We examine the 'spatial morphology' of stages for tourist performances through an analysis of the staging of the beach. Hence, both 'the edifying seashore' and 'the pleasure beach' are stages produced and negotiated through social and spatial practices. This staging of the beach not only requires reference to a rich repertoire of culturally constructed mythologies but also complex choreographies of movement.

We show this firstly, by discussing how the beach is a stage for tourism performances. Secondly, the social and cultural construction of the beach and the inscription of particular interconnected discourses on the shore of the North Sea in Denmark are examined. Thirdly, it is shown how the micro-geographical space of the beach is staged through particular practices *and* relations of people and objects. Finally, we discuss some contradictions between hegemonic and marginalized performances of the beach, and how these are entangled in the spatio-temporal rhythms of the particular beaches that are staged.

Ethnography on the Beach

To the urban cultures of the industrialized west the coastal strip where the land meets the sea is not just any kind of place. The beach plays a significant role as an attractor of fantasies and desires, one of those leisure spaces specially designated for the purpose of pleasure and physical gratification for modern city dwellers (Lefebvre, 1991, p.310). It is a site that is almost entirely distinguished through affording a space in which leisure can be performed, displayed and enjoyed. The beach is *the* emblematic space for a life of leisure. This topographical centrality of the beach in (post)modern culture is for example reflected in the significant imprint of beach life on national cultures in settler societies, such as New Zealand, Australia, or various Caribbean societies (Booth, 2001; Dutton, 1985; Sheller, 2003).

There is also a mushrooming of 'artificial beaches' around the globe. Indoor artificial beaches with simulated subtropical micro-climates have been constructed throughout the temperate zones of Europe from the late 1980s onwards. On the outskirts of major cities even larger projections of artificial beaches have been produced, such as at Reykjavik. The capital of sub-arctic Iceland now has its own beach perfected with sand flown in from the Mediterranean and waves with of artificially heated luke-warm water. The flourishing of such 'artificial' beaches point to the hybrid character of the beach in-between culture and nature.

Before the coastal zone could become such a symbolic icon it had to be constructed as a space in the first place and coded for specific purposes and practices. The intangible border zone between land and sea was transformed into a stage, a playground, a site for edification, sexual gratification or simply easy living. The coast *line* had to be (re)constructed as the beach *area*. As Shields observes, the beach 'transformed its nature into a socially defined zone appropriate for specific patterns of interaction, outside the norms of everyday behaviour, dress and activity. "Beach" became the topos of a set of interconnected discourses on pleasure and pleasurable activities – discourses and activities without which our entire conception and sense of a beach would be without meaning' (1991, p.75). Not until then could the beach become the locus for a 'liminal time-out' it historically has been in urban industrial societies (Shields, 1991, p.85). The social and cultural history of the beach made the site at the edge of the sea into a central stage for performing tourism; a performance in which 'all are actors, not spectators' (Shields, 1991, p.89). The beach is a place that only comes to life as a stage for leisure and tourism performance. The beach owes its existence to the performative aspects of tourism; aspects that generates its symbolic significance (Turner, 1985; Schechner, 1988).

Here it is necessary to consider in what sense 'the beach' can be viewed as a stage. Tourists are involved in individual and collective performances in which they are simultaneously actors and audience; performances transforming particular places, landscapes and sites into stages, stages to be gazed at or stages to step onto and be gazed at. Tourism possesses some unique analogies with dramaturgical performance. Firstly, the spaces and times of tourism are characterized by their symbolic function. Tourism is enacted within a certain symbolic framework that can be conformed to, transformed or transgressed however usually detached from any instrumental logic lying beneath the framework. Secondly, tourism usually takes place within bounded spaces; spaces that thereby are transformed into 'dramaturgical landscapes' for performing tourism (Chaney, 1993). Thirdly, like dramaturgical performances many of those bounded tourist spaces are densely managed, regulated and controlled by people and institutions acting as playwrights, directors and stage crew (Edensor, 2001).

According to Goffman performance is intrinsic to social life (1959; 1986). It is through dramaturgical actions that people create an image of who they are to other people as well as to themselves. Central to performance is a conviction that: 'a correctly staged and performed scene leads the audience to impute a self to a performed character, but this imputation – the self – is a product of a scene that comes off, and is not a cause of it' (Goffman, 1959, p.252). Identities are produced. While the essence of all performance is to present actions and identities as if they are 'real'. Dramaturgical actions are not just about producing a trickster's world of fakes and illusions. The spatiality (or 'regionalisation') of social life is intrinsically related to the dramaturgical actions of the actors.

Goffman conceives of actors as reflexive and conscious agents moving between different socio-spatial settings. Some of these settings can be understood as front stages implying intensive 'impression management' directed at a more or less exclusive audience. Other settings function as back stage regions where masks can be removed, tensions released, roles rehearsed and experiments permitted. Such front and back stage regions form a continuum of socio-spatial contexts implying different degrees of performance. Hence, Goffman claims that performances are socially negotiated not only between actors but also with a present or imagined audience in mind. 'Performances can', so he argues, 'be distinguished according to their purity, that is, according to the exclusiveness of the claim of the watchers on the activity they watch' (Goffman, 1986, p.125).

Leisure spaces like the beach are ambivalent spaces, not easily categorised as 'back stage' or 'front stage'. There is a ludic aspect in the leisure practices that transform such spaces into places to play. Places where performances are densely scripted at the same time contain an intangible element of what Turner calls the 'categorically uncategorizable'(1985). Leisure performances always contain a ludic play of subverting elements; 'an "anti" by which all other categories are destabilized' (Turner quoted in Schechner, 1993, p.25). According to performance theorist and director Eugeneo Barba this ludic aspect of performance is intrinsically linked to how bodies are used. 'We use our bodies in substantially different ways in daily life and performance situations. In everyday life we have a body technique that has been conditioned by our culture, our social status, our profession. But in a performance situation, the body is used in a totally different way' (Barba 1986, p.115). In performances the body is used as a sign, a sign to communicate culturally and socially coded knowledge transforming spaces into front-stages (see chapter 6). In this sense, the body, its appearance and capabilities are inscribed with cultural and social discourses (Butler, 1993).

The body is however also central in less discursively structured practices. In many tourist practices the experience of the body and its sensations in non-mundane practices and environments are crucial. Tourist spaces are used as back-stages to experiment with and test the body and its capabilities. Hence, there is the mushrooming of extreme sports, hedonistic carnivals or more conventional mass 'sand, sex and sun' tourism.

Throughout history the beach has been positioned between 'back stage' and 'front stage'. This is demonstrated be the clash between 19th century bourgeois ritualisation of the beach as a back-stage (strictly regulated beaches, non-mixed sexes, prescriptions for length and practice of sea bathing) and the scandals that haunted it with (peeping gentlemen, riots and systematised infidelity; see Shields, 1991; Lencek and Bosker, 1998).

The same goes for the battlefield of bodies of the current pleasure beach. As Ahmed argues, the icon of the tanned body associated with beach life displays how skin has become fetishised and obtained a function as 'a form of mask or adornment rather than holding the subject in place' (1998, p.27). The clash between the tanned bodies paraded on the beach and the rows of pink human flesh along its edge demonstrate this interplay and interchangeability of actors and audience, gazer and gazees, in a game in which the human body is transformed into a sign (Ahmed, 1998; Butler, 1993). In that sense, the performance of 'the pleasure beach' is a cultural practice positioned in a complex interplay between grotesque and idealised bodies. The dynamics of these practices transforming particular sites such as the beach into symbolically significant leisure spaces are closely associated with the interplay between conformity and transgression of 'scripts' drawn upon.

However, densely scripted and staged beach practices may seem it is important to emphasize the material and social processes underlying these performances. The staging of the beach is a social process in which some practices gains hegemony while others are marginalized. The numerous mutations of 'the beach' throughout the history of Western culture (terrifying, edifying, medical, pleasure giving, erotic and so on, see Lencek and Bosker, 1998) are embedded in an interplay between hegemonic and marginalized practices and the particular stagings resulting from it. According to Lefebvre it is precisely the destabilising and carnivalesque qualities that explain the central position of leisure spaces in contemporary society. The rhythms and routines of everyday life in the industrialised world are only possible if occasionally disturbed by breaches or carnivalsesque ruptures. The 'beach' is one of those leisure spaces that afford possibilities for rupture, for festival, for play: 'Specific needs have specific objects. Desire on the other hand has no particular object, except for a space where it has full play: a beach, a place of festivity, the space of a dream.' (Lefebvre, 1991, p.353; and Shields, 1991). The beach is such a 'space of a dream' – a field of play, an area designated for the full play of all desires.

Thus in contrast to strictly regulated and bounded leisure spaces such as the theme park or the heritage site, the beach appears as an abundant and empty space. Relatively few signs and markers instruct visitors how to behave, what to see and what routes to walk. Normally no tour guides appear to tell what to notice, what to neglect, what prescribed path to follow and where to stay at a distance. The beach is a tourist stage where tourists have to turn themselves into performers as well as a stage crew. It is a stage produced and reproduced by tourists themselves, guided by discursive knowledge, practical embodied norms and reflexive knowledge of what to do.

This chapter aims to bring out the practices and discourses enacted within the micro-geographical spaces of the beach. Such an ethnographic approach to beach life focus as on the interplay between marginalized and hegemonic practices, by examining the practices of tourists on the beach as well as the social, cultural and material resources they draw upon. And we explore how these practices enact, transform, play up against and transgress these discursive blueprints.

The fieldwork was undertaken along the North Sea coastline of Northern Jutland. Historically this region has played a significant role in generating cultural representations of the sea and the beach. It is also the most significant tourist seaside region in Denmark. Hence, the region has been central both to the formation of artisan colonies and to seaside resorts and second-home tourism.

Two particular beaches were selected for in-depth fieldwork: Torup Strand and Blokhus. In tourist guides the first is scripted as a well-kept secret, offering a glimpse into one of the last examples of on-shore fishing in Denmark. The latter is one of the best known pleasure beaches in the country. Both beaches were observed for two weeks. We conducted interviews with local shopkeepers, those fishing, bar personnel as well as tourists at different times and located in different parts of the beaches in question. Twenty interviews were conducted. The fieldwork also consisted of visual observations at each site.

Setting the Scene

The thin strip of sand separating land from the salty waters of the sea has always been a zone of ambiguity and unrest. Within Judaeo-Christian culture the mere view of the sea was a constant reminder of the flood send by God to punish humanity. The staging of the beach results from a long contradictory history of discovering the coastal zone, inscribing it with meaning and disciplining and regulating its spaces and people (Corbin, 1994).

This process took off later in Denmark than in Britain as the Netherlands. However, it followed some of the same patterns. The first coastal resort in Denmark was established on the then Danish, island of Føhr in 1819. It was in the latter part of the century that the landscapes and peoples of the seaside were discovered as leisure landscapes for the Danish urban upper-classes and henceforth became objects of their gaze. This 'discovery' was formed through a strict dichotomy between the (modern, hectic) urban life and the (authentic, tranquil) rural relics (see also chapter 5).

Particularly the extension of railroads, regular carriage and steamship connections and the automobile generated flows of affluent city dwellers to the coastal regions of western Jutland and Bornholm. This appropriation of the beach, its landscapes and its people was not simply a question of expropriating the areas of the beach and its hinterlands, but an active transformation of the beach and its people *into* objects of leisurely consumption.

In this section we show how the emergence of place mythologies assigned the beach either as spectacle or a field of play, we examine how such mythologies scripted appropriate beach behaviour as still found in current tourist practice. The Danish author Jens Kruuse recalls how he experienced the encounter with local people in his childhood holidays on the North Sea coast in the years 1912-14. In his eyes the fishermen on the shore, were remarkable different from the appearance of the men of his family and social circles. They were:

...men of another category than ours, with their pretty nice blue shirts, red ties, light jackets and straw hats. We could not talk to them. We just stood gaping at their huge fists, deep wrinkles, their faces seasoned by the salt, and their ruthless wielding of the ropes, nets, tar and canvas (Kruuse, 1966, p.105).

This account bears witness to how, the seaside was domesticated by the urban upperclass throughout northern Europe. As Corbin argues this domestication was associated with a long process of disciplination and regulation as the beach was made into a leisure zone. The beach was emptied of subverting local practices, and local people were 'tamed' and transformed into objects for the tourist gaze and servants (Corbin, 1994, p.232).

Decisive in this process was the work of artists and intellectuals. In the 1860s an artisan colony was formed in the northernmost village of the North Sea Coast, Skagen. The paintings of the sea, the shore, its people and its qualities made by this group of artists were transported into the public domain through exhibitions, urban newspapers, journals and guide books (Lübbren, 2001, 2003). This group of artisans 'made the beach into a stage on which the collision of the elements unfolded' (Corbin, 1994, p.165). However they did not only represent the beach as a stage for the edifying wonders of Nature, they also scripted and choreographed it as a stage to gaze at and to step into.

The painters of the Danish beach epitomise the ambivalence between two different paradigms in European seascape painting (Lübbren, 2003). The first is a premodern, 'grey' paradigm the second a leisure oriented 'sunny' paradigm. According to Lübbren, the turn of 19th century saw a shift from the grey, essentially romantic, paradigm of beach representations, to the 'sunny' paradigm. Both paradigms constituted important moments in the discursive scripting of appropriate beach practices. Thus, the appropriation of the shore and its transformation into 'the beach' followed two different yet interrelated lines of representation. Here we discuss these two lines of representations through paintings by the central figure of the 'Skagen school', Michael Ancher (see Figure 4.1). Both were painted in Skagen during the latter part of 19th century and became instructive in how to conceive and practise beach life.

In Ancher's *Fishermen putting a Boat into the Sea* (1881) the coast is hardly recognizable as 'a beach'. It only offers a space on which the heroic and hardworking fishermen are struggling to conquer the waves and make a living. The fishermen are imagined and visualised as people of the sea, standing in close relation to the wonders and gifts of nature. Hence, the painting illustrates how local people and their work became recast as mute creatures, yet blessed by Nature, to be appropriated by the gaze of the educated spectator. This shore is an edifying site. It is a stage to gaze at and enjoy the ongoing drama. The visitor finds reassurance and reconciliation in the powers of God and Nature, through the encounter with nature, its forces and its creatures.

The second painting presents a different scripting of the beach. *Sea Promenade* (1896) shows the beach staged as a dramaturgical landscape. The coastline can be seen in its full length, providing a setting for the women of the urban upperclasses

4.1.1   Michael Ancher: *Fishermen putting a Boat into the Sea* (1881), courtesy of Skagen Museum

4.1.2   Michael Ancher: *Sea Promenade* (1896), courtesy of Skagen Museum

Figure 4.1 Paintings by Michael Ancher

placed as the central motif of the painting. The space of the beach is emptied for all social life except for the women strolling along the shore, talking informally while occasionally glancing at the waves. The beach is made into a stage to step into. There is no distance between the viewer and the viewed. One is invited to join the scene and stroll along.

The pictures exemplify two different ways of domesticating the beach. They show how the Danish North Sea coast was inscribed with cultural meanings that prefigured how tourists would fantasize, imagine and use the beach as a leisure zone. Forty years later one can recognise the two scripts in the account of the beach given in the first official national brochure on *Seaside Resorts in Denmark* (1927):

> On the deeply cut fjords and wide bays, and on the many sounds and belts winding between a wealth of islands lie the old towns, rich in historical memorials, with their modern industrial and trade activities, and their venerable buildings, the little cosy red-roofed fiord towns; and finally, the innumerable fishing hamlets with their picturesque thatched roof sheltered at the foot of a slope on the fringe of a wood reflected in the water below.
>
> Everywhere in Denmark one meets this typical coast picture, which may be said to be symbolic of the character of the people themselves, the essential features of which are energy and sound practical sense, combined with a love of nature, and an accompanying inclination towards indulgence and gentle reveries.
>
> Everywhere in Denmark the sea makes its exhilarating and invigorating influence felt. As the light of the sun blends, as it where, whit the glittering reflections of the sea, the air likewise obtains its clearness and freshness from the same source, and gives the people their healthy colour. The sea may be said to be Denmark's cradle, its source of renewal and its song. Just as it thunders and roars over the reefs of West-Jutland and on the jagged cliffs of Bornholm, so it murmurs softly lightly lapping the gentle coasts of Funen and the surrounding islands (Rostock, 1927, p 6).

The seascape, the hamlets, the woods reflected in the water constitute a 'typical coast picture' that is 'symbolic to the character of the people themselves'. In all their diverse forms the Danish seascapes point to the 'exhilaration and invigorating influence' that induces 'clearness and freshness' (and 'a healthy colour') on the people. The sea epitomises all that is good, natural and healthy. It is 'the cradle of Denmark'. The sea, the coast, its people and hamlets are all taken to be emblematic symbols for the sound, healthy and unspoiled virtues of leading a natural life. The shore is a place where one is confronted with the spectacles of nature and its creatures (including people).

However, we also find another conception of the virtues of the seaside in the brochure:

> It is not merely the extent of the coastline, which, in proportion to the size of the country, must be said to be very considerable, but the nature of the coast itself is highly favourable for bathing. The beach along the greater part of the coast consists mainly of sand ridges, which slope gently down to the water's edge, thus creating ideal bathing conditions and permitting bathers to disport themselves to their hearts delight in the blue waves, without any element of danger. And within this sunlit playground on the margin

of the shore and the flat stretch of sand, stand either high brinks with cool shady woods
– as in many places on the islands – or high barriers of fantastically shaped dunes with
sea grass and deep sandy hollows, as is the case, for instance, along the whole of the
west coast of Jutland from Fanoe to Skagen (Rostock, 1927, p 4).

While the first description depicted the coast line as a visual scenery to confront,
gaze at and contemplate, this second one invites the reader to join in. 'The nature of
the coast itself is highly favourable for bathing...'. It creates 'ideal bathing
conditions' constituting 'a sunlit playground on the margin of the shore'. Hence, the
shore, the beach and the sun are not only components of a picturesque spectacle but
a dramaturgical landscape surrounding the playground of the beach and the
performances and bodily pleasures taking place within it.

The first national tourist brochure continues the two lines of representation also
present in the work of the Skagen school: the beach as an edifying spectacle and as
playground. It shows the way the scripts anticipated in the work by the Skagen
School and its 'lay followers' were stabilised and popularised as scripts for tourist
performances. Thus early in 20th century the scene was set and the scripts for
performing the beach were written. We see in the next two sections that these scripts
function as regulators of tourist performances and contribute not only to the cultural
representation of the beach, but to the formation and social regulation of the roles
and choreographies that stage the beach.

Performing the Edifying Beach

Torup Strand (literally: Torup Beach) was 'discovered' at the turn of the 19th
century. In the early decades of this century it experienced the first inflow of upper
class visitors. Until 1936, when the first (and only) hotel was built, visitors rented
rooms in local houses. As most other villages were developing rapidly into seaside
resorts with hotels, camping sites and second-home construction, Torup Strand
remained somewhat distanced from these developments. Plans for second-home
construction, recreational centres and so on were made during the 1960s and 1970s
but due to heavy local resistance they were not realized. The result is that Torup
Strand today is relatively 'free' of tourism developments. It is the largest, still
active, hamlet for in-shore fishing in Denmark (fifty professional fishermen,
between fifteen and twenty boats). At the same time the hamlet is a central
attraction depicted on numerous postcards and in guidebooks it is highly
recommended. The national route for scenic driving (the 'Margueritte route') and
the national bicycle route cut through the village along the shore.

The hamlet of Torup can be approached from inland as well as from the road
along the coast (see Figure 4.2). In both cases, the last mile runs westwards to the
beach with small grass-covered roads leading along the coast to the scattered
second-homes hiding in the lyme grass covered dunes. Just before arriving one
passes a small parking area. To approach the beach and the hamlet it is necessary to
continue straight ahead through a narrow passage between the dunes. The passage

is surveyed by 'idle' (possibly retired) fishers who vigorously communicate on the bench outside a small house. Standing in the middle of this passage the spectator has a beautiful view of the hamlet with the ten to fifteen boats lying on the sandy shore, nicely framed by a groyne to the right and breakwaters to the left.

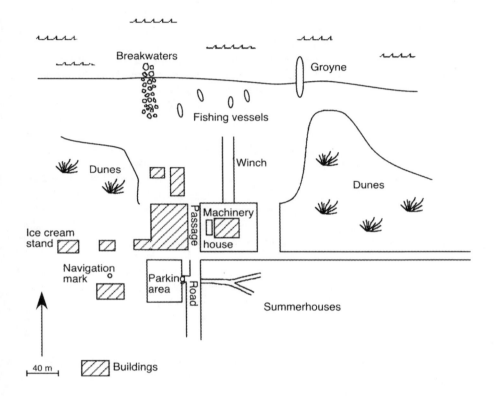

Figure 4.2 Map of the hamlet and the beach, Torup Strand

During the day the parking area is busy. Cars arrive but only stay for 15 to 45 minutes. This is just enough time to get out of the car, to stretch one's legs, go down the passage to the shore, walk around a little and take a few snapshots before leaving. Torup Strand has one thing to offer: a nice view and that's about it. Asked about reasons for coming here a Norwegian family father provides the following answer:

Interviewer:	What are you here for?
Interviwee:	We have no reason for being here. We're just passing through…

Most of the visitors are families who are 'just passing through'. They are taking a break from the scheduled program of the day. Unfortunately the ice cream stand is closed most of the day so families seeking refuge from frustrated children are quickly on the move again after having gazed at the beach, taken a few snapshots and perhaps collected some stones. It is the spectatorial 'tourist gaze' that directs, orders and to some extent produces Torup Strand as a tourist site (Urry, 2002). The beach is kept under surveillance of the eye, an eye positioned and viewing from the passage leading to the shore. Within the scene monitored, behaviour follow a strictly choreographed and regulated pattern conforming to the norms appropriate at museums too. It is a stage to look at, not to step into. Approaching this stage the spatial order of the site is evident. Framed by the groyne and the breakwaters the view afforded resembles the edifying beach of the Skagen school. It is a place to admire the alleged authentic, primitive industrial life of these coastal villages still wrestling with the sea. Stepping onto the beach and walking around is like being in an outdoor museum. The working fishermen, their boats and tools are treated as exhibition cases by the gaping visitors. Tourists move quietly around alone or in small groups (see Figure 4.3.1). Talking softly, without raising voices, hushing disobedient or impatient children, if not removing from the site to the coastline where they are told to collect stones and mussels. The edifying beach is a site of awe and contemplation best enjoyed at a distance. It is a place to gaze or take photographs. Most tourists move quickly through the area with fishing vessels, and position themselves in corners of the stage to look at and photograph the scene (Figure 4.3.2). No one talks to the working fishermen since they are visual objects. Not only do people only reluctantly engage in talk (for example, about the price of fresh fish), they also avoid looking at the fishermen while touching the vessels.

When asked about this, an elderly fisherman said that this was often a matter of amusement. Visitors behaved as if they were moving into a back stage region of local people. This view was not shared by the fishermen themselves who are fully aware of the function of the hamlet as a tourist site. As MacCannell (1976) points out, such transgressions are associated with shame. And shame and uneasiness were what was expressed by the visitors when approached by the interviewer, other tourists or locals. With downcast eyes they would flee to the corners of the site, quickly take their photos and leave as quickly as possible.

The densely regulated and choreographed practices of gazing, photographing and contemplating a site is dominant. However, they are not the only practices. Outside the margins afforded by the breakwaters and the groyne a different tourist life unfolds. Young people (locals, and second-home residents) play in the waves or just dwell (Figure 4.3.3). Also here the embarrassment of being a tourist is evident. However a German man with his family, is spending the afternoon in and around a tent raised on the beach some fifty metres from the breakwaters, marking the border between the hamlet and the rest of the beach strip. The interviewer approaches this family with general questions about their previous holiday experiences in this region, but is immediately told:

4.3.1 Moving around in groups

4.3.2 Photographing from a distance

4.3.3 Dwelling at the margins

Figure 4.3 Staging the edifying beach, Torup Strand

Interviewee:	...We are not really tourists. We've got friends here. We've come back each year since 1974...
Interviewer:	...but how did you become acquainted with these friends?
Interviewee:	Oh, it's the family who rented out their second-home the first time we went here. Then we met...
Interviewer:	Where's the house located?
Interviewee:	About 7 miles inland...
Interviewer:	But do you come here often?
Interviewee:	Every day...
Interviewer:	But... Why do you then go to *this* site...I mean the beach is rather stony and it's quite steep...Wouldn't it be more convenient to go to the beach at Blokhus. You have two children?
Interviewee:	Mmm... I don't have any opinion of that...But we like it here... because of the people... We even speak the language...

It is important for this interviewee to challenge the image of 'a tourist family' imputed by the interviewer. 'We are not tourists. We've got friends here' (...) 'We like it here...because of the people'. These statements express an opposition to the role of the tourist, and he adds 'we even speak the language'. Like this he positions himself and his family not only in geographical distance but symbolically distant from the practices of those 'tourists' (see Buzard, 1993, on such distinctions). Asked once more about what he finds interesting about coming to this site, he adds:

Interviewee:	...We just like to be here...to have a good time here together...
Interviewer:	But why do you find this particular place interesting?
Interviewee:	It's because of the people. Especially the family that rented us the house of course. But also the milieux here...it means something to us.

To this family the importance of the particular site is not its qualities as picturesque or romantic. To the contrary, they express distaste for the 'sightseeing tourists' running around between the vessels at the hamlet. What counts is the familiarity of 'the milieux here... it means something to us'. In a very concrete manner they make themselves at home on the beach using their tent to domesticate it (see chapters 6 and 7).

The family discussed here is a more extreme examples of this domestication of the margins of the beach. Also, other families took the sand road along the coast, parking in the dunes and sneaking through them. They made themselves at home by hiding in the lyme grass. On the northern margins groups of youngsters similarly found their way through the dunes for a quick cold swim from the stony beach.

While the hegemony of the eye clearly regulates the practices and performances of the stage, the sideshows display a multiplicity of marginal practices. These marginal spaces are particular populated by visitors who do not (or do not want to) conform to the role as archetypical 'tourists' but are more engaged in appropriating and domesticating the beach 'for themselves'. The beach is regionalized into spaces of play and spaces of contemplation. It draws upon the discourses of both the beach as a place for contemplation and education as well as a social space and a playground. The spatialisation of these discourses is rooted in the choreography and embodied practice of the tourists who visit the place. While this spatiality of the beach inscribes rules and norms for appropriate behaviour this is only produced through the enactment of particular discourses.

Performing the Pleasure Beach

Blokhus is the oldest seaside resort on the Northern part of the Danish North Sea coast and the first place where sea bathing became common. In the 19th century Blokhus was a favourite place for intellectuals, painters and affluent city dwelling families to go in the summer. The 'tourist encounter' between the city dwellers and local fishing culture took place in Blokhus. Since then the village and the shore have changed dramatically. During the heyday of mass tourism to the seaside from the 1930s onwards, tourist developments rapidly expand around the hamlet with large areas of second-home residences, beach hotels and camping sites. Since the late 1970s there have been continuous warnings about jeopardising the whole region because of the uncontrolled flows of tourism and associated development of facilities. Only a little more than 300 people live in the village all year.

In many respects Blokhus (and the neighbouring village Løkken) shows the Danish version of 'the pleasure beach'. Since the beginning of the century bicycle and car races have been arranged on the sands of the beach. Also today regular events such as races or sandcastle festivals bring people together. The beach is wide and possible to access by car (see Figure 4.5). It is possible to drive cars in several lines on the beach for over 15 miles.

It is also as a 'Danish Ibiza' that these two villages have gained their recent notoriety. During the summer of 2000 and 2001 parents in the Nordic countries Sweden and Norway were warned by the Chief Constable responsible for the resorts that they should be regarded a no-go area for families with teenage children.

In the summer Blokhus is a mix of teenage groups and families from the many second-homes. Compared with other seaside resorts Blokhus has a very gentle beach. The slope of the shore only declines gradually from the dunes to the sea providing a very broad beach with fine granulated white sand, low water depth up to hundred metres from the coast and gentle waves. From the hands of nature Blokhus seems a paradise for children.

To approach this beach one has to drive through the village of Blokhus – a village turned into a leisure bubble with many bars, restaurants, discotheques, gambling houses, entertainment centres and so on (see Figure 4.4). As most visitors arrive by car and there is only one road leading to the beach everyone has to queue up on the road through the village to get to the beach. The passage is surrounded by numerous pubs and bars (before having passed the dunes) and stands where one can buy ice cream, flowers and fish (after having passed the dunes).

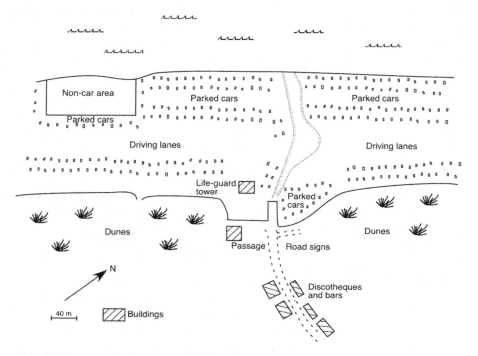

Figure 4.4 Map of the beach, Blokhus

When one has entered the beach the driving is easier since several roads run north-south along the shore. Between the lanes rows of cars are parked. On a busy day these endless rows of parked cars make the beach look more like a parking lot with a beach, than a beach onto which it is possible to drive. The traffic on the beach is even regulated by road signs. A couple of hundred metres left of the passage there is 'a car free area' into which you are not allowed to take your car. This small fenced area is more a curiosity than actually a non-car beach zone as it does not even go down to the water line, nor up to the dunes. However, it is widely used by mothers in search of a safe piece of sand for their babies or as an unguarded playground for small kids. The quality and temperature of water and wind is registered and displayed throughout the day with swimming overseen by lifeguards.

In bad weather or out of season the wide, flat beach is a desert. However on a sunny day in the peak season, as almost every day throughout the fourteen days the site was visited for fieldwork, the beach is a crowded mixture of families, cars, children, teenagers, music and so on (Figure 4.5.2). It is a chaotic mess where the road signs, the parked cars and the routes for cars seem to create some material stability to the life on the beach (Figure 4.5.1). The majority of visitors during the daytime are families with children who visit Blokhus primarily because of its friendly beach. However, these qualities are not what play a role when people are asked about reasons for going here. Hence a young Danish father (mid-30s) reflected on the reasons for going to this particular beach:

Interviewee:	It is a coincidence that explains why we are vacationing here.
Interviewer:	But why did you then in the first place choose to go here?
Interviewee:	We know the place...It's convenient for us to go here as we have been here on earlier vacations...And it makes it easier with the little one that we can put him to sleep in the shade inside the car. He has to take a nap in the middle of the day and here we can bring the car with us to the beach...

As we see 'convenience' is the word that comes to mind. To families with children Blokhus is a 'convenient beach', allowing parents shelter for small children at the same time as being able to enjoy the pleasures of beach life. A Norwegian mother (late-30s) explains:

Interviewee:	We stay at the camping site nearby...we actually go to Blokhus regularly for summer holidays...
Interviewer:	Then...Why do you come here?
Interviewee:	...Because the whole city is very friendly to children...lots of things to do, lots of facilities. We use them vehemently. I don't think there's anything particular interesting about this place...except that its friendly to children...
Interviewer:	But what about the cars...the traffic around here?
Interviewee:	I don't think the traffic is any problem... It's OK if people take care. And it's just so lovely that you can bring your own car... And then you can just sit in the car, make yourself at home and have a good time if that's what's needed...[laughter].

Again it is the convenience of bringing your car to the beach and enjoying the close-by restaurants and facilities that attracts this family. Without these elements 'there's nothing particular interesting about this place'. As in the former quotation the car is however not only a means of transport, but also a means of home-making. 'You can just sit in the car, make yourself at home and have a good time...If that's what needed'. The car is used to domesticate the beach. It is an object that gathers the family and transforms the beach into a social space for the family (Figure 4.5.1, see also chapters 6 and 7).

Though families with children dominate the beach in the daytime there are other visitors. Hence, the 'convenience beach' of the families is a stage for performance from 11am to 5pm. Before and after this period other groups dominate beach life. In the morning the beach is only sparsely populated and teenagers (single sex pairs and groups) and middle-aged couples use it for a quick bathe or as a place to enjoy their morning coffee. A couple (mid-fifties) explain:

Interviewer:	Why did you come here?
Interviewee:	We have come here for 30 years now. Since we lived in Aalborg (20 miles away). Now we have moved to (another town 60 miles south east of Blokhus)...but we still go here...now we are not able to go here except for the week-end...
Interviewer:	But why.... It's a long way to go?
Interviewee:	We like the fresh air in the morning. It's nice to have your breakfast and your coffee together with the sea breeze...
Interviewer:	But if you have come to enjoy the view and the sea breeze...doesn't the traffic and the cars make you annoyed?
Interviewee:	We like it here precisely because it's possible to drive all the way down to the water. The car is a great convenience. I think that it may be the main attraction about Blokhus...that you don't have to walk through the dunes to get to the beach... Actually...if you couldn't get down here in a car the beach would lose all its value, because the dunes are really no good...

This couple had arrived at about half past eight and when interviewed at around eleven were on their way to visit nearby friends. Although this couple explicitly mention the 'sea breeze' and since early morning had chairs arranged parallel so that they could enjoy the wide-screen seascape drinking coffee and only seldom speaking, the traffic was as they said 'no real problem for them' as it did not spoil gazing at the seascape. To the contrary they see the possibility of driving down to the coastline as a great convenience. 'It may be the main attraction' and they even go so far as to postulate that 'if you couldn't go down here in a car the beach would lose all it's value'. Although the quotation may seem loaded with ambiguities the couple do not really see any contradiction between the heavy traffic to, from and on the beach and their intention to enjoy the seascape and the breeze. In the end it is playing children, running, screaming, throwing sand in the air that causes their retreat from the beach.

When the families with children start to leave between 4 and 5pm new groups enter the scene. These are teenagers and young people preparing for the night life. Cars arrive with young men carrying football shirts, cigars sombreros streamers with 'FREE CAB, GIRLS' and the like. On the beach many beer cans are brought out for the late afternoon. Other groups (less beer, more sporty) put up nets for beach volley and shelter for their audiences (Figure 4.5.3). Whether dressed up as fake Zorro-style Mexicans or heavily engaged in performing sports the real play is not about theatre nor sports, but drinking, hunting or planning tonight's sexual 'hunt'. Mobile phones are intensely used to coordinate the gangs, the meeting places and the hunting area. As the night falls these groups move into town and leave the beach to heterogeneous population of middle-aged romantic couples, and performers of car parachuting and races along the beach. When the sun finally disappears, the beach gradually depopulates.

4.5.1 Family life in the morning

4.5.2 Cars on the beach (afternoon)

4.5.3 Teenagers preparing for the night

Figure 4.5 Staging the pleasure beach, Blokhus

While the central play performed at the beach is about the social life of families, other performances are played out in the marginal timespaces of the morning and the evening. Hence, while family life dominates the beach during daytime, it allows for a multiplicity of other plays (romantic couples, gazing viewers, teenage groups and so on). Central to all these performances is the social domestication of the beach. The beach is not staged as an object or landscape to gaze at, but as a stage for performances.

Conclusion

Tourist sites and attractions are not simply objects or a physical backcloth for tourists' consumption. Even a 'non-cultivated' and seemingly pregiven, physical space such as 'the beach' only exists as a performed space, a space inscribed with meanings enacted by tourists drawing on culturally and socially constituted discourses. Hence, 'the edifying beach' and 'the pleasure beach' discussed in this chapter, only come to life through scripted, staged and choreographed performances. The edifying beach and the pleasure beach involves people enacting discursively structured performances that have been scripted over the last century. Hence both can be traced in early tourist guides as well as to the representations of the beach in paintings.

While it is possible to identify the dominant performances of beach life at the two sites these are not unified. Rather they reflect the hegemonic practices inscribed at the various sites. A number of marginal practices, spaces and times have been identified. These also suggest that performing tourism is not wholly a pre-programmed practice. Tourists are not cultural dupes. Performing tourism draws on culturally and historically discourses on how to appreciate sites and attractions and appropriate behaviours at such ones. However they also include reflexive knowledge. This can be seen in relation to 'the edifying beach' where it was shown how the reflexive knowledge of 'being a tourist' informs both non-communicative behaviours by visitors as well as the highly reflexive identity-work of self-proclaimed 'non-tourists'. It may also be seen in the apparently ambivalent roles of romantic gazers and mass-motorized visitors in relation to 'the pleasure beach'.

Chapter 5

Photographing Attractions

I have been photographed and knew it. Now, once I feel myself observed by the lens, everything changes: I constitute myself in the process of "posing". I instantaneously make another body for myself, I transform myself in advance into an image (Barthes, 2000, p.10).

I have taken two types of photos today. Some pure landscape pictures...I've tried to capture the beautiful landscape motifs...And, of course, then the other pictures where you picture your kids against the historical background; and...your kids in a funny situation unaware of the camera (interview 5).

Introduction

Tourism and photography are modern twins. Vacationing is the single event where most snapshot images are made, and it is almost unthinkable to travel for pleasure without bringing the lightweight camera along and returning home with many snapshot memories. Although photography is perhaps *the* emblematic tourist practice and tourist studies have been dominated by a visual paradigm (see chapter 1), tourist studies have produced little knowledge of *how* and *why* tourists are busy producing photographic images. The existing literature tends to be of a speculative nature, and it portrays tourist photography as a wasteland of pre-programmed shooting where tourists are not so much framing as already framed by the tourism industry's spectacular economy of signs (Sontag, 1977; Albers and James, 1988; Urry, 1990; Osborne, 2000; Crouch and Lübbren, 2003).

Inspired by Crang this and the following chapter *reframe* the approach of tourist studies to visual culture by stressing the sociality, creativity and embodiment of tourist photography (1997, 1999). We view such photography as a theatre where tourists perform various scripts, roles, technologies, relations and places to and for themselves and for a future audience. 'The very idea of representation', as Said observed, is a 'a theatrical one' (1995, p.63). Performance metaphors – stage, script, director, acting and so on – can help us write illuminating and dynamic accounts of tourist photography as an embodied and creative performance 'full of life' that *produces* memories, social relations and places. Thus, the *snapshot* metaphor, and its undertones of headlong shooting, is dismissed on the ground that it prevents us from registering the physicality, temporality and creativity of much photography.

Performances of tourism photography are a complex amalgam of materiality, technology, discourse and practice, by non-humans as well as humans. If by agency we understand the capacity to act or to have effects, then the camera is an agent. It can act in specific fashions – for instance freezing a moment of time on paper – and it permits certain actions and not others. The humanly performed aspects of 'ordinary' photography are visible in relation to practices of *taking* photos – discovering and framing views, exhibiting and directing people in front of the camera – and *posing* for cameras. Tourist photography performances are not enacted for the camera in a social vacuum. They are scripted by, and acted out, in response to dominant 'tourist gazes' and mythologies, that circulate in photo albums and the 'imagescapes' of television, films, magazines and so on.

Through a framework of such theatrical metaphors that combine the 'real' and 'imaginary', the material and immaterial, this chapter examines picturing practices and images. The research was conducted at northern Europe's largest medieval ruined castle, Hammershus, on Bornholm. Hammershus is surrounded by, and has spectacular and extensive views of, the sea, cliffs, and dales. The combined cultural and natural grandeur of the place have charmed tourists for some 150 years. It is by far Bornholm's most visited site. Everywhere one looks at this traditional high-cultural sight, there are children and families. It is an exceptional place, not to be missed when touring the island, even if one has been there before.

The chapter explores *how* sightseeing photography is choreographed *and* acted out, corporeally, materially, socially, culturally in different fashions and through different 'tourist gazes' (Urry, 1990, 2002). We examine what happens to the 'tourist gaze' when it interacts with the camera. Discussions of the 'tourist gaze' have not been sufficiently examined in relation to technology or performance. It has been shown that photography corporealizes the 'tourist gaze': it transforms consuming, distanced spectators into active directors and actors who constantly produce new 'realities'. The study brings out how different photography performances produce different 'attractions' within the same place. In particular, it examines how tourist photographers perform attractions as sights *and* as stages for acting out social life, in the latter case providing a picture of how, and to what extent, tourism's attractions are transformed into theatres for staging a 'nice' family life.

We develop the notion of the 'family gaze' to capture 'family-orientated' photography. The 'family gaze' introduces questions of sociality and social relations into discussions of tourist vision and photography (see also Haldrup and Larsen, forthcoming; Larsen, 2003; Chambers 2003). It differs from other 'tourist gazes' by producing photographs that first and foremost capture and celebrate family members and friends. The family gaze revolves around the *production* of social relations rather than the consumption of places. Material places are not unimportant to this vision; rather, it performs places differently from the other gazes. Places become scenes for acting out and framing active and tender family life for the camera. Family members and their performances make experiences and places extraordinary and full of enjoyable life. This is what Wearing and Wearing (1996) would call a 'feminised gaze' since it stresses interactions, relationships and the actively embodied use of space. This vision makes fluid the boundaries between the ordinary and extraordinary, between everyday life and tourist life.

The notion of the 'family gaze' enables us to connect the sociological literature on 'intimacy' (Giddens, 1991; Beck and Beck-Gernsheim, 1995; Smart and Neale, 1999), with that on family photography and tourism (Chalfen, 1987; Bourdieu, 1990; Kuhn, 1995; Barthes, 2000; Holland, 2001; Chambers, 2003). Tourist photography is one of the modern ways in which families produce life-narratives that construct them as families in a mobile world. When performing tourist photography, families are creating experiences and performing their 'familyness'. Photography is part of the theatre that modern people enact to produce their *desired* togetherness, wholeness and intimacy (Hirsch, 1997, p.7; Smith, L. 1998, p.16). In this perspective, tourist photography becomes a drama of acting-out and capturing 'perfect' family life in an era of 'family fragments' (Smart and Neale, 1999). Family images are thus never simple records of 'real' family life, but are shot through, consciously and unconsciously, with desires, fantasies and ideals of family life, of 'imaginary families'. Yet we do not make a simple comparison between 'real families' and 'imaginary families' since all families are 'fictions' in the sense of created and performed.

To capture and examine tourist photographic performances, ethnographically inspired research methods are required (as argued in chapter 4). Ethnography is a method for accumulating cultural knowledge. Yet, rather than collecting knowledge already 'out there', ethnography is a process of *producing* and circulating knowledge that is bound up with the researcher's 'way of seeing' (Pink, 2001). Inspired by Martin Parr's (1995) captivating photographic work on 'snapshot-tourism', our particular way of observing photography performances is structured around photographing 'photographing tourists'. Since cameras and photographers are ubiquitous at Hammershus, people seldom noticed that our camera was aimed at them and not the attraction; they do not pose for our camera. To bring out tourists' own accounts of their photography performances, 20 semi-structured qualitative interviews were conducted at the exit from Hammershus. Except for one German couple, the interviews were with Danish tourists, and they were mainly families holidaying with their young children. These interviews lasted from ten minutes to almost half an hour (see interview list in end of chapter). To be able to examine tourists' own photographs, we asked for a copy of their Bornholm photos at the end of the interviews. Ten 'families' subsequently provided us with these photographs that 'capture' their day in that place.

While tourist performances transform places (see chapter 1), the stages in which photography takes place are always inscribed with cultural scripts and material regulations crucial in choreographing tourists' cameras. Humans (guides, guards, tourists and so on) and 'non-humans' (markers, fences, pamphlets, guidebooks, postcards and so on) exercise such choreography. Therefore, before we examine tourists' photography performances, the cultural, material and social scripting and staging of Bornholm and especially Hammershus are explored.

Scripting and Staging Hammershus

The following examines the early making of Hammershus and Bornholm as a place for tourism within travel literature and painting (see Larsen, 2003, for a more detailed account). Like the Lake District and the Swiss Alps (Urry, 1995; Ring, 2000; chapter 8 below), Bornholm's 'wild' and 'rough' landscapes and seascapes were little praised before the arrival of tourists in the nineteenth century. In the late 1790s, a young Copenhagen-born poet described it as the most horrible and frightful island on this entire earth. So awful was this wild and rocky place that he prayed that no soul would have to walk in his footsteps.

Bornholm's re-imagining gathered force around the very beginning of the nineteenth century with the publication of the topographical travel works by the Copenhagen-based P.N. Skougaard (1804) and O.J. Rawert and G. Garlieb (1815) (Skougaard was raised on Bornholm). While they belonged to a well-established tradition of *scientific* travelling aiming at meticulous surveys and objective descriptions, their texts and sketchers were certainly not without poetic dimensions and an eye for the aesthetic. To quote Rawert and Garlieb:

> Such landscapes, such powerful and clear-cut formations are very few, and none are to be found in Denmark. Bornholm has such a great abundance of the most romantic landscapes, pleasant and delightful as well as very mountainous, frightful, gloomy and so on…that aesthetic souls and poets craving romantic landscapes…should rather travel to Bornholm than Switzerland (1815, p.56).

The last reference to Switzerland is particular striking. 'The Alps are not simply the Alps. They are a unique visual, cultural, geological and natural phenomenon, indissolubly wed to European history' (Ring, 2000, p.9). Just as the English made the Alps (Ring, 2000), Bornholm became inscribed with the sublimations of the Swiss Alps: 'Denmark's Switzerland' (Jensen, 1993). This reflects the way places come into existence through *relationships.* They move around within global networks of humans, technologies, imaginations and images that situate them within a cultural economy of difference *and* similarity (see Hetherington, 1997).

Many of the first pioneer tourists to roam Bornholm were city-dwelling painters who flocked to the countryside across Europe (Lübbren, 2001). Rawert's portrayal of it as a Swiss panorama of picturesque beauty and sublime drama triggered off an immense desire for Danish artists to sketch this 'foreign' place in their own country. A local paper wrote in 1857: 'This year artists are visiting our island in unprecedented numbers, recording views at different places, and in Copenhagen a considerable number of today's prominent painters have exhibited landscape studies and folk culture motifs from Bornholm' (in Jensen, 1993, p.52, our translation). This traffic initiated the 'mechanical reproduction' that is crucial in the making of tourism places (MacCannell, 1976). It created, sustained and disseminated the place-myth of Bornholm.

Situated gracefully in the middle of this 'Swiss' scenery on an elevated spot, Hammershus became one favoured scene for visiting painters. Picturesque painters had long popularized ruins and historical monuments and 'a ruin in a proper state of decay' was a favourite sight among tourists in England (Andrews, 1989, 1999; Ousby, 1990). Yet Hammershus was in an unhealthy state of decay before it was preserved as a historical monument beginning in 1822. For half a century locals had consumed it as site for collecting stones and bricks. For the eighteenth-century inhabitants of Bornholm, the deserted castle apparently represented little more than a valuable assortment of useful building materials (rather like Hadrian's Wall in northern England). The castle gave birth to many new houses, with the result that its former potency was disappearing. 'Outsiders' and romantics considered that it was disgraceful that such a significant national symbol and charming picturesque monument was being ruined by 'vandalism', in peacetime and by Danish hands, into the bargain. They initiated the preservation of the castle and years later picturesque painters restored its romantic grandeur.

In 1849, Anton Kieldrup bathed Hammershus in such a picturesque light and perspective that one feels transported to a timeless idealized Italy (Figure 5.1.1). The whole picture corresponds to a Claudean picturesque schema of the beautiful and theatrical with its perfect harmony between nature and people. For a start, the painting is composed of three distinct zones – foreground, middle ground and background – which meticulously frame all the elements like sets in a theatre. The foreground is the darkest, yet also the most richly detailed part. Rocks, on the one side, and the little native herd boy sitting in the grass, on the other, frame and guide one's vision along the path into the lighter middle ground – which is in turn framed by ancient trees – where a 'theatre of painting' literally unfolds. Through the painters' bodies and gazes, the spectator's eyes are guided directly to a castle on the horizon that is portrayed in a gentle and slightly dreamy state. The painting has an extraordinary atmosphere of serenity, timelessness and romanticism. The painters included in the middle part are significant, as they authenticate the place's picturesque nature. Kieldrup's desire was to beautify the place, to compose it as picturesque, to make it a suitable object for tourist sketching and gazing.

The sublime soul longed for the powerful and dramatic impressions that nature's great upheavals produce: raging waterfalls, rocky coasts and lashing seas. Holger Drachmann's *The Coast South of Hammershus* (1870) is an illustrative example of how Hammershus was also painted within a framework of 'the picturesque sublime' (Figure 5.1.2). In the foreground, the sea is raging and waves are dramatically thrown high in the air as they clash with the mighty, almost vertical cliffs. The cliffs themselves are portrayed with sombre darkness and their completely barren surfaces instil a feeling of inhospitality and danger. The low, heavy clouds further add to this atmosphere. Yet in the midst of the dark skies a fissure allows gentle sunrays to penetrate through to bathe the castle in a romantic light; the ruin lies majestically with its entire solid grace in a sea of the raging sublime. The whole 'landscape' emanates the emotional geography of pleasant fear that 'nature's' sublime theatres were supposed to invoke in the distanced spectator. Performing the place with pencil on canvas in the styles of the picturesque and the sublime respectively, Kieldrup and Drachmann conjured up and displayed Hammershus as a proper ruined castle for the 'romantic gaze'. Such

5.1.1   Anton E. Kieldrup: *Hammershus* (1849), courtesy of Bornholm Art
         Museum

5.1.2   Holger Drachmann: *The Coast South of Hammershus* (1870), courtesy of
         Bornholm Art Museum

Figure 5.1 Hammershus paintings by Kieldrup and Drachmann

5.2.1   *The West Coast with Hammershus and Hammeren*, photo: Lars Gornitzka

5.2.2   *Hammershus Ruins seen from the East*, Colberg Publishers

Figure 5.2 Postcards of Hammershus

paintings were instrumental in attracting tourists as well as choreographing how tourists were to consume this special place.

This brings us to the multi-layered – cultural, material and social – choreography that takes place at the attraction *today*. Figure 5.2 consists of typical contemporary postcards of Hammershus, which are on sale at the attraction as well as throughout the island. Their resemblance to the above-analysed paintings produced more than a hundred years earlier is striking. They reflect how much contemporary tourism marketing imagery is rooted in a visual language already developed in the eighteenth and nineteenth centuries. Again we see how Hammershus and its surrounding landscapes are portrayed as a timeless, awe-inspiring and pristine attraction with a 'life of its own'. To sustain this 'place-myth', the many tourists are erased. Most of the postcards are devoid of human beings while a few portray lonely crowds in the distance. Not only is the place exhibited as a proper ruin and landscape for the 'romantic gaze'; it is choreographed as a place exclusively offering visual consumption. Other tourist activities such as having a picnic, playing ball games, sunbathing are not pictured and are therefore discouraged (see Crawshaw and Urry, 1997, on 'people' in 'landscape photographs').

Cultural choreography also takes place through signs. Hammershus is staged with numerous signs, what MacCannell (1976) calls 'on-markers', which through texts and rich illustrations 'guide' tourists through the dramatic history of the castle. They provide tourists with a script for visualizing the place as a thrilling and heroic mediaeval castle where 'hanging', 'torture', 'prisoner escapes', 'excessive indulgence' (drinking, eating), 'bloody battles' and so on took place. Their work is to stimulate the tourists' imagination and make them see what is now invisible. They clearly cater for men that would delight in consuming brutalities, battles and death (see Lennon and Foley, 2000, on various 'dark tourisms').

Yet a marked 'conservationism' towards historical objects is evident at Hammershus. This 'authentic' building is not wrapped in, or overshadowed by, 'postmodern' gimmicks. There is no theming or Disneyfication making spectatorial and even mediatized gazing possible. Only on very special occasions is entertainment on display. There are no tools, technologies or activities that enable children (and adults) to taste, touch, hear and see the place through active bodily engagement. One cannot dress as a Hammershus warrior, photograph 'medieval' soldiers and so on. While the markers cater for the imagination, it is people's own task to conjure up the dramatic events that have taken place there. The small and visually 'dull' museum is tellingly separated from the 'authentic' castle itself. It is placed next to the cafeteria from which it takes around five minutes to walk to the castle.

Omnipresent signs dictate that crawling, climbing or just playing around on the ruins are prohibited; thus effectively proclaiming that leisurely walking along the clearly demarcated pathways is the only appropriate form of movement. By prohibiting climbing, the authorities attempt to prevent unnecessary *physical* contact between tourists and the ruins. While the other senses are not catered for, turning the castle into a wonderful view-producing machine for the 'tourist gaze' enhances the visual sensations at Hammershus. As a towering fortress it has had a powerful 'military gaze' looking out for enemies built into it from the start: today it is tourists' cameras that do the shooting. This reflects the fact that the 'tourist gaze' is not only a way of seeing and representing landscape, but also a way of making landscape material so that it presents itself as a nicely framed 'picture'.

Figure 5.3 provides an illuminating picture of the main viewing spaces. The first image effectively brings out the fact that the Hammershus benches typically have a view. They are viewing-stations where tourists, in the comfort of the 'chair', can consume laid-out panoramic and imposing sights (5.3.1). Figure 5.3.3 and especially Figure 5.3.4 show how the other major viewing-stations have been created by work upon the ruins: a wall is levelled out; a bridge, stairs and a security wall are put in place to make walking and gazing possible and safe. If we compare this with the adjacent piece of wall, it is evident that work and materials have been invested in producing this viewing-platform for tourist gazing. Yet by not being 'polished of', it appears authentic, rather than man-made for tourists. Interestingly, some of the viewing platforms turn their backs on the castle itself, and all of them face the 'cliffscape' and 'seascape'. Hammershus is not only an object of the 'tourist gaze'; it also produces views.

An old-fashioned sightseeing-boat tours the coast of Hammershus regularly during the summer. Out here at sea level, tourists have a low, open and distanced view that nicely amplifies the beauty and grandeur of the rock. One is face to face with the rocks, literally looking up to them (see Figure 5.4.2). Through such material choreographing, the beauty of Hammershus comes into its own. The benches, the viewing-stations and the sightseeing-boat produce distanced views that unfold and exhibit the castle and its landscapes as a picturesque theatre. They allow tourists to produce romantic photos without investing much work or skill in it.

Hammershus is remarkably little policed by 'external' forces or powerful surveillance mechanisms such as CCTV cameras and personnel are *not* in place. Because few guided tours are on offer and organized sightseeing parties rarely tour the castle, tourist performances are not dictated by the rigid and repetitive *team* choreography of tour guides (see Edensor, 1998). The responsibility of producing and sustaining social order and guarding the vulnerable ruins is assigned to 'non-humans' – signs and paths – and to tourists themselves. People comply with the unwritten decree that a historical ruin-landscape requires, and indeed deserves, respectful behaviour. Practically no adults are seen 'mountaineering' the fortress. One does not pee or drink or drop litter or sunbathe at Hammershus. It is famously clean, immaculate and well mannered. The social tone and dress code are very casual, and showing off, 'alternative' behaviour and groups of youths are out of place. The normalizing, self-regulating gaze of such middlebrow tourists makes external regulation almost superfluous, signs are enough.

We have shown how nineteenth-century painters and writers constructed Hammershus as a much-celebrated space of picturesque sights. The contemporary tourist industry sustains that place-myth through brochures and postcards as well as by working upon the castle materially. The assembly of viewing stations, benches and so on turn it into a machine producing views of romantic scenery.

5.3.1

5.3.2

Figure 5.3 Viewing-stations at Hammershus

5.3.3

5.3.4

Figure 5.3 (continued)

Hammershus is choreographed as a space for the visual consumption – gazing and picturing, of romantic nature and authentic ruins, while tourists are expected to become aloof and semi-detached spectators, almost as if in a museum. Despite being dominated by families with young children it is *still* choreographed primarily for 'romantic gazing'. With the slight exception of the markers, no work is put into children or adults who are young at heart with added thrills. No activities allow people to experience Hammershus through bodily enactment – to become an 'actor'. Other practices are subtlety discouraged, even though the open spaces of the castle invite many diverse activities. Drinks and food are *not* on sale at the castle *itself*, and tables that would provide space for prolonged socializing and food consumption do not accompany the many (view-producing) benches. One is clearly not invited to 'stay for lunch'. In this sense, the romantic postcards reflect material and cultural choreography. Now that this multi-layered scripting and staging has been outlined, the remainder of this chapter explores the photographic performances that take place.

Photographic Performances

Virtually all tourists arrive at the 'official' entrance of Hammershus by car or bicycle with their partners, children and friends. (Some people arrive on foot at the back from the shore. These few are predominantly keen walkers.) Along the laid-out paths people meander, gaze and photograph in terms of their own particular body rhythms, movements and improvisations. Extremely few couples or families tour the site without at least one camera. While all the cameras at Hammershus were in 'action', only a minority were 'overheated'. The typical visit – for families as well as couples – lasts from 45 minutes to an hour (the time spent at the museum and cafeteria is not included) and most groups produce five to ten pictures. Few families take more than twenty pictures, although several use more than one camera (the figures are roughly the same for the few families using digital cameras). Few can be said to have experienced the place only through a lens (see Edensor, 1998, pp.128-35, for 'tourist life through a lens' at the Taj Mahal).

Video-filming tourists are more inclined to 'frame' their experiences. Typically they film for 10-15 minutes, but some manage twenty minutes despite staying for less than an hour. While many families are equipped with both an ordinary camera and a video camera, some travel with a video camera alone, but these are few in number compared with ordinary cameras.

Just under an hour is enough time to walk the main paths at a leisurely pace, to read the signposts, and to make stops at the major viewing stations to gaze and photograph. Few tourists stay longer than an hour and produce more than 20 images. More pictures would probably be taken if shows were enacted and tourists could dress up as warriors and so on. Many people have Hammershus photos from earlier visits so this is another explanation of the relatively modest number of pictures taken. Yet despite having a collection of more than 900 images, one man with a keen interest in photography took around 20 on that day (Interview 18).

The research showed that most tourists have little interest in the technical aspects of photography. Flashy, arty cameras are far outnumbered by, fully automatic 'pocket' cameras. Furthermore, manual cameras are often set to 'automatic'. Light and handy, designed for strolling and extremely user-friendly, the automatic 'pocket' camera is the emblematic family camera, used by both adults and children. 'The one we have chosen to bring along today is one of those snapshot cameras, an Olympus, because it's so easy to take pictures with it for the family. It's a superb camera for sightseeing and things like that...We also have one of these fancy manual cameras, which we use in other contexts. But not when we are going places like this; then it weighs too much: it's too awkward to bring along' (Interview 16).

Freed from the 'boring', technical side of photography, tourists take pleasure in choreographing 'social world' and displaying 'material world' for the recording camera. Tourists invest time and creativity in producing pleasing images with a personal touch. Both in words and actions, people express their joy in making pictures, in experimenting with composition, depth, choice of motif and not least in having fun directing and posing. 'When we're video-filming we tend to do everything twice, because if it was fun the first time, we have to do it again for the camera' (Interview 2). Most regard tourist photography as a pleasure rather than a burden, as an integral to pleasurable and memorable sightseeing. One mother travelling with her extended family stated:

> It's incredibly pleasant to take photographs, and that's the reason why I have this kind of camera. It's so easy to take pictures with; you capture so many amusing incidents...and I wouldn't dream of taking pictures if it felt like a duty...Then I'd rather not bother, because that feeling would be obvious from pictures when I looked at them afterwards...For me, pictures are definitely things that come to life because a desire to see it again (Interview 16).

Children too delight in photography. They often set up performances and energize them, both as performers and photographers. Many children kids have their own colourful and toy-like (disposable) cameras. Over and over again, one could hear children begging for camera action, and parents explaining that loaded cameras were not cheap playthings.

The observations, interviews and photos collected indicate that *two* main ways of photographing are found at Hammershus. These are the 'romantic gaze' and the 'family gaze'. As one father says: 'I've taken two types of photos today. Some pure landscape pictures...I've tried to capture the beautiful landscape motifs...And then of course the other pictures where you photograph your kids against the historical background; and...your kids in a funny situation where they're unaware of the camera' (Interview 5). Another father says, 'I think we document two things...we document the ruins, their history and charm...And we document our own presence here – that is, family documentation' (Interview 9). While most tourists perform *both* photographic visions, they tend to favour one over the other. Different experiences and images are produced within the same space. Hammershus is not a single place. The rest of this chapter examines how tourists perform two gazes, bodily, socially and technically.

The Romantic Gaze

Most of the interviewees articulate a desire to capture pleasing images of the ruins and landscape of Hammershus. The viewing stations discussed above are powerful in attracting and immobilizing tourist flows. Virtually no one misses out on such open views; walking is put on hold, eyes gaze, fingers press shutters, bodies pose and loud travel talk is heard: 'Wow! What a wonderful view!'. The viewing stations prompt and choreograph the cameras of tourists – men, women and children – to make images that reproduce the enduring place-image of Hammershus as picturesque romantic scenery. Nine out of ten families produced a 'dead ringer' of Figure 5.4.1 (compare with 5.3.2) - four of them had photographed it from sea level (Figure 5.4.2).

Yet most tourists state that they only take a few such romantic or 'postcard' photos partly because they are difficult and time-consuming.

> Speaking from experience, gazing at a landscape...is a totally overwhelming experience: you want to have a picture of it. But when you come home, it isn't so overwhelming any more...such photos are often very disappointing...The ruins of Hammershus? Well I've taken a few, just to have a little reminder. I can't focus it down and capture the whole place properly (Interview 6).

People struggle to take proper possession of the vast landscapes; they know that the actual views from the viewing-platforms and sightseeing-boat are richer and fuller than the pictures that they are capable of making. This resonates with Pocock's findings from Durham Cathedral (1982). While photography reproduces nature magically, the small 'reproduction' is not necessarily magical. From experience people have learned that it takes more than just pressing the shutter button to produce enchanting landscape images. Skill and an eye for composition are required. This explains why most people, even at the viewing-stations, are more eager to praise the scenery with travel-talk, than to click with the camera's shutter.

The problem of capturing the wholeness of places with the camera explains why video-filming tourists are keen to scan landscapes by meticulously moving the recording camera slowly from one side of the body to the other. In the hands of most tourists, camcorders are superior to still cameras in taking 'possession' of vast scenery ('possessive gaze': see Berger, 1972). While the handheld camcorder allows tourists to integrate their walking practices into the filming, to become a walking human camera, in practice they tend to shoot with their feet at rest, static and fixed. They only move their upper bodies. Both cameras thus *im*mobilize tourists' mobile vision in practice, gazing-while-walking. They are not photographing while walking. This gives shape to the experience: stop, take a photograph and move on. Strolling and picturing rarely go together.

The people who photographically perform Hammershus through the 'romantic gaze' are the relatively few with *artistic* aspirations and sophisticated manual cameras. They like to experiment and be creative, and some of them have a technical interest in cameras too. They are the 're-embodiment' of the nineteenth-century tourist who gazed at and sketched nature artistically with 'fine art' aspirations (Andrews, 1989). To quote a German girl travelling with her partner: 'Yeah,

5.4.1

5.4.2

Figure 5.4 Tourists' photos from Hammershus

definitely, I take much pleasure in making photographs...I'm always looking out for views and pictures...I'm already thinking about the photo-book we can make when we return home...[I invest] lots of creativity and work in it...and money too' (Interview 10). And as a mother says, 'I can walk around for a very long time, looking for *the* view...and *the* proper composition. I like to try out different angles to get it perfect...To compose the picture a little. I can spend quite some time on that' (Interview 8).

Such 'arty' tourist photographers also stress the challenging work of making pleasing images, of taking possession of vast landscapes through the camera. 'Of course, very tricky...it's very difficult to get the right spot, the right time and...the sun is not out' (Interview 10). Instead of giving up, they make this their defining pursuit: taming, capturing nature's beauty and beautifying it in a small image on paper. The young man with the awesome collection of Hammershus images describes the fascination of photography in the following words:

> The fascination of making photographs? That is, well, the difficult part is...that landscapes are enormous when you gaze at them, right? If you can capture it and nail it down on a little piece of paper...that is the greatest...With a camera you can also capture the ambience of the place visited and how it 'touched you' – how you felt... . [With] the right light a picture can be made that is superior to the actual experience – if you're really lucky and invest a lot of time. It may also be that you've taken some photos of Hammershus, say, four years ago, some really nice ones where the castle stands majestically in the sun. Years later you realize ... that something has changed about the 'motif'; a spire is broken, for instance. Then you have a unique picture and a new motif. That's pretty awesome (Interview 18).

In typical picturesque fashion, the fascinating power of photography is described as beating nature – partly through magic, partly through skill – at its own game of producing its charms. Such photographers strive to produce, not just pleasing pictures of Hammershus, but unique ones. This also explains why this particular photographer constantly finds new 'pictures' within a space that he has pictured so extensively over the years. He notices and pictures the smallest alterations in the make-up of the ruins from year to year.

The 'romantic gaze' generates a desire for untouched scenery and solitary viewing so that it can realise pleasing images. While the first tourists had few problems gazing and photographing like this, at the *now* hectic and noisy Hammershus where other tourists and young families always surround one, it takes much work and patience. In such a lively environment it is arduous to experience romantic tranquility or sublime awe. While most people express little annoyance over the numerous tourists, it bothers these photographers greatly: 'if there weren't so many people, it would be a better place' (Interview 10). The photographer with the many Hammershus photos complains – with a smile on his face – that tourists disturb his artistic work: 'Well, that's the thing with the tourists again, right? Well, if you're lying on your stomach you have to be careful that your head isn't stepped on by a kid or something like that' (Interview 18).

For such a vision, the art becomes that of 'recording over' other people. The ideal is a picture where the power and charm of Hammershus glow in timeless, majestic solitude. To achieve this, patience is crucial. To quote the German woman again, 'I always spend like 15 minutes waiting so that there are fewer people...so I don't have all these horrible tourists in my pictures' (Interview 10). People are seen as trash, an eyesore. The work of such photographers revolves around not letting tourists spoil their images and memories. It is in the future space of the photo that the anticipated Hammershus is to unfold and materialize.

Figure 5.5 illustrates some typical 'romantic gaze' performances – captured by us. It shows that photographers indeed enact this vision in solitude even when sightseeing with 'significant others', their often-fancy camera is their only real 'better half'. This is neatly illuminated in Figure 5.5.1 where a posturing father is taking an artistic picture of a piece of wall. In his all-absorbing attention to the stones of the ruin he is blind to his family right next to him. While men and women both picture the castle through this vision, 'artistic' romantic photography is more common among men than women, and the bulky professional-looking camera remains primarily an item of male jewellery. There is a male bias to the 'romantic gaze', and it is more popular with couples touring *without* children.

It is difficult to find the time for such photography when sightseeing with small children. The picture reveals that the father is doing his artwork while the family is having a small break at a bench. 'The great thing about travelling is encountering all the new motifs...then you can really have fun with your camera. Yet you rarely have ten minutes to just lie flat on your stomach and experiment with your scene and camera. It bores your relatives to death: they're eager to move on' (Interview 18). The two keenest 'arty' romantic photographers interviewed were emblematically travelling only with their partners. But this is not unproblematic either. Both companions had little interest in (such) photography and they had to put up with their partners' time-consuming photography.

Figure 5.5 also illustrates some of the physical exertions that people indulge in to produce proper shots: bodies erect and kneeling, bodies bending sideways, forward and backward, bodies leaning on ruins, bodies hanging over cliffs, bodies lying on the ground and so on. Yet while the bodies are clearly active and move around, the photographic event still immobilizes the photographer by bringing their movement to a halt. The photographer pictures while at 'rest'. Squatting is particularly common, reflecting the idea that the low view is picturesque: a squatting viewpoint amplifies the magnitude of Hammershus. Figure 5.5.2 is interesting because the photographer has chosen a ground-level view, rather than an elevated one from the viewing platform just behind him. Other traditional picturesque strategies and styles are documented in Figure 5.5.3, where the ruin is framed within a 'window', while 5.5.4 demonstrates that such photographers construct 'depth' in their pictures, in this case, by 'flowering' the foreground. Such photographers are not only producing images of Hammershus, they are also using it as an outdoor studio for experimental photography.

Figure 5.6 consists of some 'arty', 'romantic gaze' images produced by two different photographers. Experienced photographers with a practical knowledge of composing and framing techniques, as well as a desire to create extraordinary holiday images, produced these images. Rather than the common shots from the viewing platforms, these are 'unusual' views. The most striking is 5.6.1. Black-and-white

5.5.1

5.5.2

Figure 5.5 Photographic 'romantic gaze' performances

5.5.3

5.5.4

Figure 5.5 (continued)

photos are certainly 'out of place' in the joyful and colourful universe of holiday imagery, it belongs to the world of art. Yet here it is put to use with dramatic effect; it underscores perfectly the dramatic staging of the raging sea and the muscular cliffs. In the midst of all the sunny, delightful family life, the sublime is captured. The duskiness of black-and-white films is sublime production *per se*. Partly because they are in colour, his other photos (5.6.2) have a lighter and more romantic touch.

The two long-distance pictures of the castle (5.6.3, 5.6.4) are also eye-catching. Although taken from different angles and by different tourists, they are almost identical. Both are taken from a point outside the 'beaten track' of paths, signs, children and so on. The two photographers have moved 'outside' the place and in traditional picturesque fashion are looking *in* with a distanced view. The fence and tree respectively make this clear, as well as creating depth in the pictures. The classic three-layer perspective is employed to perfection. It is when the tourists distance themselves from the crowd, moving their bodies to the margin of the place where silence reigns and nothing obstructs their 'camera eye', that Hammershus performs its most enchanting 'pictures' and shows its timeless romantic charm, a haunting place.

5.6.1   5.6.2

Figure 5.6 Tourists' 'romantic gaze' photos

5.6.3

5.6.4

Figure 5.6 (continued)

These photos greatly resemble the professional 'postcard' (compare with 5.2.2). They (re)produce the Hammershus place-myth of romantic grandeur that early artistically inclined tourists invented and popularized. Here we have an example where tourist photography's so-called 'hermeneutic circle' is verified in practice. Yet, ironically, it is those tourist photographers who are most skilful, engaged and creative who lend some credence to the idea of tourist photography's 'economy of reproductions'.

The Family Gaze

For most visitors though the ideal Hammershus photo is not postcard-like. Many bring their own camera as one woman says, 'in order to take our own personal photographs, not just the postcards' (Interview 6). Making 'postcard' images has little appeal because they do not convey personal experiences and stories. They are far too *im*personal and dead. 'Don't get me wrong: the landscapes here are beautiful enough to look at, but empty landscape photos are not that attractive: they have no personal ambience' (Interview 19). For such tourists, postcard images of landscapes and attractions are extremely boring, as one mother frankly makes clear:

> Well, we've learned never to take photographs without people in them because it's bloody boring to see a ruin without any people you know... . So we choose some motifs that we think are beautiful or have a nice view, but we make sure that the boys or one of us are in it to make it a little personal, so that it isn't just a postcard, because then we could just have bought it down in the shop, right? (Interview 12).

In the interviews, people emphasise that they desire, and reflexively attempt, to make *personal* images, and that holiday photography is *family*-oriented. Such tourist photography is a family ritual for the mobile yet inward-looking family. 'It's closely tied up with the family, right! If you're here as a young couple or as a single, then I don't think that you'll...take pictures of Hammershus or other attractions in the same way as we have...it's a family-related project' (Interview 12).

Despite praising its outstanding beauty, these tourists do not consider Hammershus a setting worth portraying for its own sake unadorned by family faces. On its own, neither the symbolic aura of the castle nor its romantic grandeur gets their cameras trembling with anticipation. 'Loved ones' have to enter the 'picture' to attract and energize the camera. They do not take photographs of the castle as such. They bring it into play as a backdrop for family staging. To quote another woman:

> No, we haven't taken photographs of Hammershus, we've taken atmosphere-pictures where the family is at the centre, you can see that it's a holiday, and Hammershus is in the background. But it [Hammershus] mustn't fill the image, it's the family that has to fill it, isn't it? And then the little reminder of where we are. That has to be in the background (Interview 13).

The desire is to make photos that connote 'holiday' and memorialise the family's communal experiences. As a recognized tourist place, Hammershus fits that role, but so do numerous other sites that are less visually spectacular, such as the place where they are staying (this is developed in the next chapter). As a photographic object, the castle holds no particular value for such families. It is just one suitable stage among many others for family photography.

However, most tourists expressed a simultaneous desire to make pictures *of* – not only *at* – Hammershus. The art of this 'family gaze' is to place one's 'loved ones' in the attraction such that both are represented aesthetically with grace. While expressing an equally blasé attitude towards *pure* landscape images, they are looking out for viewing-stations, 'beautiful spots' and 'nice views' to frame their family members *and* the attraction. The aim is here to produce *personalized* postcards, to stage the family within the attraction's socially constructed aura.

Producing a proper 'family gaze' requires work, patience and skill. For a start, as Figure 5.7 (produced by us) indicates, the 'family gaze' is often enacted at busy scenes. Not many people seem to be uncomfortable with staging the 'intimate' in the midst of the 'public'. People are not looking over their shoulders, or waiting for other tourists to get out of the line of sight, or rushing the event. We can also see that passing strangers take little interest in other people's 'private' dramas. That Hammershus is brimming with families 'on set' does not cause 'snap-rage'. 'Most people are polite enough to move when they see you with a geared-up camera. Few cause annoyance by passing straight through a setting. People have become more considerate of one another, aware that other people want good family photos too; they're very thoughtful, (Interview 8). Another father: 'I don't even think about it. It's a tourist attraction, so you have buses and lots of people. That's the reality... . Well, if we want to stage and take a picture of the family...we just wait our turn' (Interview 9). Thus a whole culture of waiting, taking turn and 'looking out' for photo performances is in place. The tourist is entitled to photograph his/her family on its own. That is why the 'family gaze' can be satisfactorily performed in a crowded place like Hammershus. The photo performance pictured in Figure 5.7.1 is a telling example of this. People wait patiently for the family to make their picture before moving for their turn.

That the 'family gaze' is enacted through the active participation of *all* family members is also brought out in Figure 5.7. The family is both the subject and the object of the photographic event, and everyone seems to fulfil both roles – picturing and acting. Children – even very young ones – are often behind the camera. Parents appear much more eager to portray their children than their spouses, and if the children did not reach out for the camera too, holiday memories would be less crowded with adult faces. Children are crucial to the working of the 'family gaze'. Collaboration is necessary in order to portray the desired united family and to create a complete travelogue. As one photographing mother complained to her posing husband: 'Can I be part of the holiday as well? Take a picture of me with the kids and the ruins too' (overheard conversation). Often families ask other tourists to take a photograph of them so the whole family can be portrayed in the same picture.

Yet it seems that women are more passionate 'family gazers' than men are. They take more responsibility for, and pleasure in, creating domestic memory-stories. In that sense, this vision accords less with the apparent male dominance of the 'romantic gaze' and the masculine basis of the 'tourist gaze' (as argued by Veijola and Jokinnen, 1994; Wearing and Wearing, 1996; Morgan and Pritchard, 1999).

5.7.1

5.7.2

Figure 5.7 Photographic 'family gaze' performances

5.7.3

5.7.4

Figure 5.7 (continued)

Figure 5.8 Choreographing tourist photography

Figure 5.8 **(continued)**

The 'family gaze' produces photographic events typified by dense *corporeal* and *social* performances – acting, posing, directing and so on. Walking around and observing photo action people also seem to inhabit another body. Activities and walking are put on hold and in posing people present themselves as an idealized future memory. In accordance with the late modern cultural code that tenderness and intimacy epitomize blissful family life, so families act out tenderness and intimacy for the camera and one another. Indeed, where the 'family gaze' holds sway, nothing appears more natural than producing moments of tenderness and intimacy. The 'natural' is a cultural code that has to be performed before it can be represented as real (Bourdieu, 1990). This vision supports Beck and Beck-Gernsheim's (1995) contention that modern family life is fuelled by a frenzy of love (see next chapter).

In particular, such 'intimate geographies' are produced by codified performances of visual and corporeal proximity of embraces and eye contact. Figure 5.7 brings out the fact that touch – body-to-body – is essential to the 'family gaze' when cameras appear, almost as a reflex they assume tender, desexualized postures holding hands, hugging and embracing. 'Arms around shoulders' is *the* common way of bonding family members as *one* social body.

Figure 5.7.2 is a telling example of this. Standing just behind his son and wife, a father *connects* the three family members by moving his head a little forward and putting his arms around their upper bodies. This masks his physical supremacy, and he is not 'overshadowing' the wife and son. The tender and unified family ideal is produced by exhibiting the family on an 'unequal footing' (literally, we have a family body with three heads and four legs). When adults are taking photographs of, or posing with, children they compensate for disparities by kneeling down – typically to the level where their heads meet those of the children. Again we see how the desired model is the emotionally unified family, and for that ideal to come into being signs of physical and symbolic inequality must be obliterated.

Body-to-body experiences cause, and indeed are a *sign* of, unprecedented moments of intimacy and love. 'Touch is above all the most intimate sense, limited by the reach of the body, and it is the most reciprocal of the senses, for to touch is to be touched' (Rodaway, 1994, p.41). Touch acts directly upon the body. Therefore, being touched can also be an embarrassing experience, repulsive even. Glimpses of fathers and especially teenagers who appear slightly uneasy and even resist embraces were also observed on several occasions.

The collage also brings out how the 'family gaze' produces images that are arranged around responding eyes. Human eyes and the camera eye act in responsive reciprocity. When confronted with the camera most people automatically face it. Few defy the camera eye by turning their faces away. The human eye and the hybridised camera eye are face to face. Nothing disturbs such proximity. Again we see how photographers, in order not to enact a 'belittling' gaze, bend their knees. The ideal seems to be direct eye contact, if not a view from slightly below. This creates a viewing position in which the viewer of the image 'looks up' at the exhibited family. Relentlessly exhibiting face-to-face proximity, the 'family gaze' produces extraordinary events *and* images of intimacy and co-presence. Simmel argues that one cannot take through the eye without at the same time giving (Frisby and Featherstone,

1997, p.112). The future spectator is face to face with those photographed and this produces commitment and involvement (and see chapter 6).

The way people display their bodies is not solely in their own hands. They are subjected to efforts at choreography. All the time, you can hear and see photographers attempting to stage proper faces and bodies. Expressions like 'Smile!', 'Say cheese!', 'Look like you're on holiday', 'Don't fool around', 'Don't make faces', 'Look into the camera', 'Move closer', 'Put your arms around each other' and so on are common and often complemented by energetic body language.

Clear examples of the choreographing enthusiasm of photographers are revealed in Figure 5.8 (images produced by us). Twelve successive images shot at intervals of between three and ten seconds showing two women's determined efforts to stage and capture their four children. First, we witness the staging of the event. As if ill-clad for camera work, perhaps feeling too hot and stuffy, the camera-wearing woman takes off her jacket. Then, meticulously, one after another, she positions the boys (only the older one is taking his own seat). Next, the actual shooting begins. She squats so that the 'camera eye' is more level with the eyes of the children. Direct eye contact is established. Now the other woman joins in the action. Standing just behind the kneeling photographer with her eyes fixed on the boys, she gesticulates vigorously with her arms in the air. Then a small break occurs and the photographer changes shooting position, straightening up her body slightly. Now events intensify. For the next minute or so the photographer constantly frames and shoots, while the other woman's arms make all sorts of disco-aerobic moves and shakes – all acted out with a big smile on her face. Although the boys' arms are not 'joining in', their faces are probably laughing and a joyful holiday photo gets produced. While excessively choreographed, it is emblematic in the sense that 'family gaze' photos are rarely the outcome of a quick shutter release. They are enacted, embodied lengthy performances.

This also shows how most performances in front of camera are static. Posing, for instance, is motionless – at most there may be waving and so on. More a result of social conventions than technical necessity, vigorous action is rarely captured. Paster (1996) has compellingly argued that the performances are 'still' because a desire for timelessness fuels photography, immobility yields immortality (see next chapter).

However, Paster's argument is too reductive. Many tourists are also fascinated with the video camera because it affords mobile recordings 'full of life'. In many families, a division of labour between the two technologies is in place: 'with the still camera we picture dead things, like the castle here; the camcorder is used when people are 'on' and moving about, to record some life' (Interview 4). That the camcorder permits the recording of *sound* – the waves of the sea, family chattering and so on – was mentioned as crucial in conveying an ambient atmosphere and personal narratives. The 'family gaze' celebrates life and the camcorder is clearly at home in such family life.

Tourist photography produces unusual moments of intimate co-presence rare outside the camera's eye. The proximity comes about because the camera event draws people together, symbolically as well as corporeally. Tourist photography simultaneously produces and displays the family's closeness. To produce signs of a loving, intimate family life, families need to enact it bodily. Family frictions are put

5.9.1

5.9.2

Figure 5.9 Tourists' 'family gaze' photos

5.9.3   5.9.4

Figure 5.9 (continued)

on hold when the camera appears. Stressed parents, bored teenagers and crying kids are instructed to put on a happy face and embrace one other before the camera begins to click. In this sense, it is cameras, public places and cultural scripts that make *proper* family life possible – relaxed and intimate. However, they do not create the 'proper' family. Photography is not something that transcends and negates reality; it creates performances in which people work to establish realities (Crang, 1997, p.362). The desired family is the product of the photographic event that each family stages and performs actively and bodily. It is the enactment that produces 'familyness' (see also chapter 1). In other words, photography performances produce rather than reflect family life. Yet they reflect – document – the produced closeness and naturalize – render invisible – the artful nature of the image.

We can explore this further by examining emblematic Hammershus images that tourists have produced. The interviews suggested two versions of the 'family gaze'. The first way of picturing families demonstrates little affection for the castle (Figure 5.9). These are pictures *made at, not of Hammershus*. In some of them it is in fact difficult to see that they were taken there at all. Clearly it is not the castle's aura or attractiveness, but joyful moments of socializing, playing and posing that trigger off the camera: close-ups. They could in principle have been shot anywhere. Figure 5.9.1 is particularly interesting because the photographer captures the 'family gaze' in action: a candid close-up video-filming event played out between a father and his

5.10.1

5.10.2

Figure 5.10 Tourists' 'romantic, family gaze' photos

daughter. Although they are 'sightseeing' at a much-celebrated cultural sight, both filmmakers only picture family bodies. In this way the castle is actively used as a stage upon which embodied family stories are played out for the camera. The ruins of the castle exist in order to frame the family. The sight is overshadowed by the father filming the daughter, with the mother photographing the photo session between these two. Hammershus 'disappears' in these photos. This renders the 'seeing of sights' insignificant; picturesque greatness and order turn into a misshapen and indiscriminate assortment of stones, benches, lawns, humans and so forth. As the romantic legacy is undercut, so it actively contributes to the production of a new place-myth of a joyful and playful family life. This collage exemplifies how tourist photographers not only consume and reproduce long-standing myths and postcard realities but also inscribe places with fresh cultural meanings.

The pictures in Figure 5.10 were pictures *of* the attraction. They were taken from a distance, and the depth perspective allows the grandeur of the place to be taken in and the family to be fixed in the aura of the attraction. Such pictures are effectively a hybrid of the romantic and family gazes. Figure 5.10.1 is choreographed such that the father and two daughters are not upstaging the attraction. The photographing mother has created a symbiosis between the two. The image has a well-composed balance between being-there and being-together. These pictures work by being classic tourist snaps that capture an exceptional encounter between a family and an acclaimed, extraordinary attraction. They are postcard-like. Yet the placing of family members in the picture inscribes endlessly reproduced sights/images with personal aura and meanings.

Again we can see the solemnity of tourist photography. People show their respect for the photographic event and their social relations by posing in a dignified manner. Dignified and polite poses are endlessly performed at Hammershus; it has become the 'natural' way of displaying oneself, and posing itself has become a natural performance. Particular salient is the absence of so-called 'post-tourist' performances of irony, whether directed at Hammershus or family photography. Seriousness typifies the 'lightness' of this small world of tourist photography.

Conclusion

This chapter has shown how tourism photography amounts to more than just a 'way of seeing'. Conventionally, photography is portrayed as a static, distanced and disembodied encounter with the world. But it has been shown here that corporeal work, creativity and the other senses are involved in performing proper tourist photography – whether in front or behind the camera. Touch, talking and body language are crucial to the production of holiday images.

There is more to sightseeing than seeing a site. Picturing practices are an integral part of sightseeing. Almost all tourists combine the consuming 'tourist gaze' with the *productive* camera. Performing photography is not simply a way of documenting pre-existing experiences at an attraction, but part of producing them as concrete bodily performances and tangible memories. Not only do many tourists derive pleasure from performing photography in itself; properly staged images also ensure, no matter how insignificant, boring or disappointing the actual experience was, that the desired atmosphere will be projected into the future. Rather than portraying sightseeing as consumption of fixed spectacles on distanced screens and the sightseer as a spectator, we have seen how cameras and attractions offer access to a stage for 'lay-artistic' experimentation and story-making, for directing and acting.

And there is more to sightseeing photography than capturing a site. Even at somewhere like Hammershus scripted and staged for 'romantic gazing', the making of photos is significantly bound up with and revolves around social relations. For many tourists, Hammershus first becomes an attraction for their camera when 'loved ones' take the stage and enact joyful 'familyness'. This is not necessarily an indication of people's lack of interest in, or disappointment with Hammershus. Rather, it reflects how photography and tourism are major social practices through which people in the contemporary image-saturated world produce storied biographies and memories that make sense of their lives and social relationships.

The notion of the 'family gaze' is central here, formed in opposition to the 'romantic gaze' and to the public, crowded 'collective gaze'. This family gaze deploys places as stages for framing personal stories revolving around social relations. The identity of tourist photography is formed in mobile space between home and away, between extraordinary places and events *and* family faces. The distinctions between the private and the public are blurred in family photos: people perform the 'private' with 'public' scripts. This produces what Sheller and Urry have called 'hybrids of private-in-public' (2003, pp.115-17). It reflects how tourist and family photography are thoroughly wedded to each other. In a world of organized mobility, tourism is not an 'exotic island' but is highly interconnected with ordinary social life. In the next chapter we examine in more detail the social basis of the 'family gaze' within the particular context of late-modern family life and its memory work.

List of Interviews conducted at Hammershus (chapter 5 and 6)

(1)   Middle-aged parents with two daughters aged 7 and 13 respectively.
(2)   Two Danish families, middle-aged parents with young children.
(3)   Elderly couple travelling on their own.
(4)   Family with two young children. Only the parents participated in the interview.
(5)   Interview with a father travelling with his wife and two children aged around 8 and 10.
(6)   Interview with a father and his daughter-in-law travelling with their extended family.
(7)   A father and his son around 10-12.
(8)   Family with parents and two children.
(9)   Teenage daughter, a younger brother and their middle-aged parents.
(10)   German couple in their late twenties travelling on their own.
(11)   Parents in their late thirties with two young children.
(12)   Middle-aged couple with two young children.
(13)   Couple in their late thirties with two boys aged around 6 and 7.
(14)   Couple with two young teenage daughters.
(15)   Interview with a father vacationing with his family.
(16)   Couple travelling with teenage children. The latter did not participate.
(17)   Family consisting of two adults and two children.
(18)   Couple in the mid-twenties travelling on their own.
(19)   Couple in their thirties with two boys around 8 and 12.
(20)   Couple with two young teenage daughters.

Chapter 6

Memory Work

"How sad it is!" murmured Dorian Gray, with his eyes still fixed upon his own portrait. "How sad it is! I shall grow old, and horrible, and dreadful. But this picture will remain always young. It will never be older than this particular day of June ... If it were only the other way! If it were I who was to be always young, and the picture that was to grow old! For that – for that – I would give everything! ... I would give my soul for that! ... It will mock me some day – mock me horribly!" (Oscar Wilde, 1951, p.31).

Introduction

Sontag famously states that tourism 'is a strategy for the accumulation of photographs' (1977, p.9). While there is clearly more to tourism than taking photos, tourist photography is concerned with accumulation. The previous chapter showed that most tourists invest time in, and derive pleasure from, the actual work of photography. The desire to capture delightful memories in image form animates peoples' photography. In this sense, each sightseeing 'events are not so much experienced in itself but for its future memory' (Crang, 1997, p.366). Although some tourists considered holiday photography time-consuming, slightly boring and even a little embarrassing, people still brought the 'annoying' camera with them and frequently used it.

This chapter argues that people are drawn to and engage in tourist photography in order to accumulate memory-stories. It explores why tourists are so busy accumulating memories and why the production of memories assumes an important role. Tourist studies have been slow to recognize that the social importance of holidaymaking exceeds beyond the period of the 'two-week holiday' (chapter 1). Through photography practices, people strive to make fleeting experiences a lasting part of their life-narrative.

One way in which people make sense of themselves and their relationships is through narratives (Giddens, 1991, 1992; Shotter, 1993; Gergen, 1994). Tourist photography is part of a theatre of narratives and memories. Rather than an alienating superfluity, photography is an integral component in the production of identity and social relations. We might even suggest that 'people require [tourist] adventures in order for satisfactory life stories to be constructed and maintained' (Scheibe, 1986, p.131). Becoming a tourist involves exceptionally high expectations of a distinct time, a time when people are most excited about life and their bodies are geared to pleasure.

Here we examine how tourists use photography to construct memory-visions reflecting idealized family and holiday life. This chapter explores the social basis of tourist photography within the context of late-modern 'family life'. It discusses why the 'family gaze' is crucial to social relations of family intimacy.

The chapter is partly based upon interviews conducted at Hammershus (see chapter 5), and partly on analysing nearly 1000 photographs produced by Danish and German holidaymakers in Bornholm and Jammerbugten, North Jutland (see also chapters 2, 4 and 7). Inspired by Lutz and Collins (1993), we apply a dual analytical strategy to the photos collected. A 'content analysis' is the first step. From the many collected images a quantitative 'picture' is extracted to provide a systematic idea of the places, events and people that are materialized as tangible memories. This demonstrates how a wide range of places are captured and transformed into theatrical spaces for photography, and shows that more than half the images portray family members and friends 'performing places' or acting for cameras.

Secondly, five emblematic 'family gaze' images are culturally analysed. This latter analysis sheds further light on and explains the dominant cultural *codes* – the implicit collective rules – governing and constituting the 'family gaze'.

Finally, the complex production of memories through photographs is analysed. What do photographic memories do for people and how do they display, value and work upon them – physically and mentally? It is shown that 'family gaze' photography revolves around the fixing of elusive but blissful family moments. The desire to stop time animates such photography. While photos fix time, they are not receptacles for memories. Photographic memory is a complex of relational interactions between humans and images. People's active engagement with and work on images stimulates contingent, short-lived memories.

Small World

This section analyses the content of the 'photographic world' produced by the families on holiday. Since photography is performed, and since certain activities provide better photo events than others, such an analysis cannot 'document' tourists' experiences or evaluations of the two places in a simple sense. However, it does provide evidence of the tangible memories that the 21 families produced and took home with them. With regard to the actual coding, two of us spent two days practising until a >90 per cent correspondence with random images in the coding procedures was reached. Subsequently, test samples were taken with the same results.

Table 6.1 shows the distribution of the 'stages' or scenes captured in the 937 photos analysed. Since tourist vision has historically been bound up with the idea of landscape as scenery (Andrews, 1989; Ousby, 1993), it is not surprising that 'rural landscape' is most frequently pictured, accounting for about a quarter of the images. Given that tourist studies have long argued that the consumption of attractions is vital in tourism (MacCannell, 1976), and that both destinations (Bornholm in particular) have well-known cultural sights, the relatively low number of such pictures is striking (14 per cent). A wide range of 'stages' are employed and several of them are mundane and not typical objects of the 'tourist gaze'. More strikingly, 'residence' comes second in the distribution; more photos are taken at the holiday home than at attractions. The 'beach' is also popular (14 per cent), while cars, ferries, bicycles and restaurants are common. In particular, the popularity of the home, the emblematic material object of the 'family gaze' as a photographic stage, indicates the significance holidaymakers ascribe to tourism as a 'way of being together' (see chapter 1).

This also raises the question of the 'acting dimension' of the pictures. More than half of the photographs contain one or more family members or friends in the foreground. This indicates the popularity of the 'family gaze'. Such strong emphasis on family faces and bodies should be seen in relation to the strikingly low number of other 'tourists' and 'locals'. Other people are deliberately choreographed out of the public, crowded, places. The eyes of the camera avoid other tourists. Holidaymakers desire 'private' photos. The people of northern Jutland and Bornholm are not viewed as a 'picturesque other'.

Hence, substantially fewer than half the pictures show deserted 'stages'; scenes worthy of portrayal on their own. While reproduced in postcards and brochures, the classic tourist image, a romantic picture of a deserted 'cultural sight' or 'rural landscape', only accounts for a quarter of all the photos (yet it is worth noting that these two 'stages' in fact *are* most frequently pictured without people). The table also reveals that almost all the home images are populated. Thus, one explanation of many home images is that this 'stage' is especially appropriate for picturing family bodies. In the cosy geography of traditional 'familyness' in the holiday home or caravan, family pictures can be created. Most images of cultural sights are also populated with family faces. Such photos provide 'physical evidence' that one has been there and that many tourists find empty attraction images 'boring'. The previous chapter showed that tourists make an effort to picture a busy attraction without other tourists. Table 6.1 reveals that unfamiliar faces are not allowed to 'spoil' the foreground on any of the images. At best other tourists are allowed to inhabit the blurred background.

Examining the bodily actions of the people portrayed (in Table 6.2), we see that 40 per cent of them choreograph their bodies for the photograph. Other activities are suspended and people present themselves for future memories by 'posing'. This testifies to the 'theatrical' nature of tourist photography. 'Playing bodies' are the second most popular performance. Interestingly, adults are also pictured as playing, fooling around and showing off as juvenile bodies. The emotional sociality of family tourism is also popularly captured around the dinner table – whether in the countryside, in the second home or a restaurant. Thus various modes of performance and interaction among family members are photographed. Somewhat surprisingly at first glance, 'gazing', the motif that the tourism industry employs to represent the pleasures of the landscape, is only third most popular. Gazing is not a suitable motif for tourist photography, because it requires people to turn their backs to the camera while the 'family gaze' is based on eye contact between the family and the photographer.

Performing Tourist Places

Table 6.1 Photographic scenes
in %*   No. of pictures = 937

Sites / Actors**	Family members	Locals	Other Tourists	No people	Total
Rural landscapes	9	-	-	17	**26**
Residence	13	-	2	2	**17**
Beach	11	-	-	3	**14**
Cultural sights	11	-	-	3	**14**
Amusement parks, zoos, pool areas	7	-	1	2	**10**
Urban landscapes	7	-	-	1	**8**
Means of transport	3	-	2	1	**6**
Restaurant or other Small facility	4	-	-	-	**4**
Museums and galleries	-	-	-	1	**1**
Total	**65**	**1**	**5**	**29**	**100**

*	Only values more than 1% mentioned
**	Only actors in the foreground of the picture

Table 6.2 Staging the family
in %*   No. of pictures = 520

	Residence	Beach	Rural Landscapes	Amusement parks etc.	Sight	Café etc.	Urban Landscapes	Transport	Total
Body Display	6	6	8	4	6	1	6	3	**40**
Sport, play	7	10	2	7	2	2	-	-	**30**
Gazing Photographing	-	2	5	-	3	1	-	1	**12**
Eating, drinking	5	1	2	-	-	3	-	-	**11**
Reading, writing, games	4	-	1	-	-	-	-	-	**5**
Domestic work	2	-	-	-	-	-	-	-	**2**
Shopping	-	-	-	-	-	-	-	-	**-**
Biking, running	-	-	-	-	-	-	-	-	**-**
Total	**24**	**19**	**18**	**11**	**11**	**7**	**6**	**4**	**100**

*	Only values more than 1% mentioned

The previous chapter brought out the 'intimate geographies' produced by holiday photography. In tender ways, bodies move together and eye contact is established to present bonding relationships and to stage love, the new opiate of the masses (Beck and Beck-Gernsheim, 1995). In most images that portray 'actors', there is at least one figure in *visual proximity* to the 'camera eye'. The photographic event moves people together. There are few pictures where people are dispersed in space. While touching – embracing and hugging – is common when people pose in a group, most bond by exhibiting *corporeal proximity* within an 'intimate range' that is within touching distance but not actually touching.

Family Gaze Images

The modern nuclear family *does* exist and is flourishing as an ideal: as a symbol, discourses, powerful myth within the collective imagination. This cultural myth is a regulatory force that impacts on our lives at a very personal level (Chambers, 2001, p.1).

We now examine the social basis of the 'family gaze' and tourist photography within the context of modern family life.

Various marketing reports indicate that the birth of a baby and going on holiday are the two main reasons for purchasing cameras (Chalfen, 1987, p.75). The 'snapshot literature' has shown that for most people photography is intricately bound up with social relations, in particular with 'family life'. 'Photographic practices only exist and subsist for most of the time by virtue of their family function' (Bourdieu, 1990, p.14). In rituals of the domestic families use the camera to display success, unity and love; it is put to work to immortalize and celebrate the high points of family life. The typical image is of 'loved ones' taken on special occasions like births, weddings, parties and holidays. Mundane everyday life and strangers have no place in the photo album (Chalfen, 1987; Paster, 1996). '[You take pictures] of events you may not experience again. You take pictures of the extraordinary: when your baby is born; when you get a dog in the family and things like that. And, of course, when you go travelling' (Interview 7). Like tourism, family photography produces a 'small world' of positive extraordinariness.

The blissful family produced by family photography contrast in many ways with 'reality'. It is claimed that the family is under siege, as signalled by growing divorce rates, single parenthood, joint custody, singles, step-families and so on. The traditional nuclear family, the life-long marriage between a heterosexual man and woman and their (two) children living under the same roof, is fragmented and dispersed. Yet, the failure of the late modern family has not undermined people's faith in, or desire for, family life. To get married and have children is still 'natural', and the nuclear family is 'a powerful myth within the collective imagination'. What is new is that splitting up and remarrying have become normal too. So the remedy for the so-called crisis of the family is the family! Beck and Beck-Gernsheim argue that:

Figure 6.1 Tourist's photo: At the second-home

Figure 6.2 Tourist's photo: With the car on the beach

...What will take over from the family, that haven of domestic bliss? The family, of course! the negotiated family, the alternating family, the multiple family, new arrangements after divorce, remarriage, divorce again, new assortments from your, my, our children, our past and present families. It will be the expansion of the nuclear family and its extension in time; it will be an alliance between individuals as it has always been and it will be glorified largely because it represents a sort of refuge in the chilly environment of our affluent, impersonal, uncertain society, stripped of its traditions and scarred by all kinds of risk. Love will become more important than ever and equally impossible...This, then, is what we mean by the normal chaos of love (1995, pp.2-3).

No matter how many shattered illusions and broken hearts experienced, people strive for 'family life' because that is where love supposedly resides in an individualized, secularised modernity. People live in a frenzy of love, in the 'normal chaos of love'. The 'family' is being recovered as a 'pure relationship' in a democracy of contingent love. It is the romantic complex of 'forever' and 'one-and-only' qualities with which Giddens contrasts his notions of 'pure relationships' and 'confluent love'. 'Pure' love is lived out in 'impure' families. Such relationships exist because of love, and if they do not deliver enough emotional satisfaction, they tend to break up (Giddens, 1992). Bauman describes this as 'liquid love' (2003).

In an era of 'pure relationships', where the values and institutions that once legitimized the family and bonded it together are losing power, 'families' constantly need to perform acts and narratives to give sense, stability and love to their family relations. The more family life becomes fluid and based on choices and emotions, the more tourist photography can be expected to produce accounts of a timeless and fixed love. Photography's logic of irreversibility and non-erasability serves as the perfect medium for the modern quest for a fixed identity (Bauman, 1995). The glueing of snapshots into the chronological album literally and symbolically fixes these memories into the irreversibility of history. The photo album immobilizes the flow of family life into a fixed series of eternal snapshot identities. It reflects a sense of identity constructed by ties, long-term commitment, continuity and memory, representing what Lash and Urry call 'glacial time' (1994).

In the following, we analyse five typical 'family gaze' images in order to see how tourist photography is enacted and places are performed to produce the pleasing 'familyness' of solid love in an era of fluid 'families'. The *first* image (Figure 6.1), portraying a father and his two daughters casually and tenderly playing on the lawn of the rented holiday home, is a typical example of the 'residence' and 'playing bodies' images. In such pictures, everyday behaviours are slightly suspended or inverted into more playful, free and sensuous ways of being together with 'loved ones' and in 'material worlds'. Their attraction is the captured moments of spontaneous family joy – juvenile bodies captured in the framework of the holiday home's imaginative geography of the cosy family dwelling; a utopian place where 'ordinary' family life becomes extraordinary, relaxed, shared and joyful. As Wang puts it, tourists are not merely searching for the authenticity of the *Other*. They are also searching for authenticity in and among *themselves* (Wang, 1999, p.364). Many family holidays involve the desire to be together as one social body, to experience close social proximity in an era of fragmented family life.

'Corporeal proximity' is central to this picture. The three bodies are connected and form a 'moving train'; the father is the driver and the two girls' bodies are the coaches. As *one* familial body they 'move on'. While they seem partly aware of the photographer – the mother! – they are not posing as such, and the playing does not seem staged. The picture has an aura of 'naturalness'. The mother has crafted an image where her own presence disappears and the 'mirroring' camera captures the spontaneous moment. Ironically, the making of such 'natural' images requires the photographer to have the patience to wait for and the eye to see the decisive moment when people actually perform 'naturally' – that is, appropriately in terms of certain codes. Crying children, non-docile teenagers and stressed parents have no place in the 'candid' camera stories of families. The 'natural' is a cultural code (Bourdieu, 1990, p.166).

The *second* image (Figure 6.2) works through a similar hybridisation of 'home' and 'away' and the 'public' and the 'private'. It is also deconstructs dichotomies of what is social and natural, what are subjects and what are objects. Here the photograph is taken on a beach in northern Jutland, but the posing family bodies and the displayed car domesticate the wide-open space. While the subjects are out there in 'nature', all activity is kept close to the car. For instance, the older sister – who appears to be unaware of the photographer, sunbathes at touching distance from it. The family inhabits the car – a mobile home – and the car inhabits the beach, creating a private space in the public one. The sea and beach are upstaged.

Touching the tailboard, the father shows his pride in his car and pays it respect for making a proper family and holiday possible. They are flexible and mobile. The mother has choreographed an image portraying the car as an indispensable travelling companion. Paradigmatically, the father and the daughter maintain eye contact with the photographing mother. The viewer is guided directly into face-to-face contact with them and the car. The image reaches out to the viewer; it stares him/her in the face and engages. Recorded visual proximity produces intimacy and co-presence.

The *third* image (Figure 6.3) differs from the previous two by portraying two families holidaying together. Thus the 'family gaze' is creating 'family' by enacting intimate relations with significant others. The content analysis showed that friends are commonly captured too. The 'family gaze' is not exclusively bound up with a traditional notion of the 'family', whether the nuclear or extended family. It pictures and produces friendships too. Shot on the beach, the image portrays a relaxed 'bodyscape' of children and adults. The semi-naked bodies and the work that these bodies have produced in unison – the sandcastle – make the picture. The sandcastle is surrounded and framed by their bodies, and the girl in the foreground guides the viewer's vision straight to it. Pictured thus, the sculpture becomes a monument to friendship between the two families and across age boundaries. This communal project leaves a fragile, short-lived trace in the sand that the recording camera eternalizes by turning it into a tangible, exchangeable memory. The intimacy of the image is also produced by the exposed yet desexualised bodies. Clearly they feel at ease exhibiting their semi-naked bodies in each other's photo albums. This is in itself a symbol of the intimacy and pureness of their friendship. Interestingly, it also shows that it is not only children who are pictured as 'playing', fooling around and showing off. The images collected here show many adult bodies as 'juvenile bodies'.

Figure 6.3 Tourist's photo: The sandcastle

Yet despite the image's aura of naturalness and playfulness, we see that the adults closest to the front make slightly different bodies for themselves; in accordance with the code that slim bodies are healthy and aesthetically pleasing, they breathe in a little. When people sit for portrait photos they already imagine themselves as an idealized memory before the shutter button has been pressed. They present themselves as a future image. Furthermore, there is something distinctively solemn about the picture. Everyone – except the little girl – expresses respect for the photographic event by posing in a dignified way; gentle smiles are worn, bodies are straightened, hands are kept at sides. No one pokes fun or monopolizes attention. Although they are not touching one other, their almost identical poses produce *one* social body that is ceremoniously displayed. Like the scenes at Hammershus, such dignified poses are performed in the images collected. This is the 'natural' way of displaying oneself, and since people pose facing the camera, posing itself has become a 'natural' performance. Photographic poses are not 'natural', yet in creating perfect bodies and moments they appear so.

Figure 6.4 Tourist's photo: Picnic on the rock

Figure 6.5 Tourist's photo: On the bike

Figure 6.4 perfectly embodies the landscape taste of the 'family gaze'. Rather than picturing itself in a detached, aesthetic way, the family has crafted a pure image of its pleasant 'dwelling'. Nature is practiced and represented as an embodied scene for active social life and animal life. Framed by the dogs in the foreground and trees in the background, the family sits *in* the landscape, socializing on a large rock, having a picnic. Tourist photography as family memory is generally hampered by the fact that somebody has to take the pictures. This makes it impossible to picture the family as a nuclear unit. The father, the mother or one of the children is always absent – behind the camera. This family, consisting only of the four people and two dogs pictured, has solved the problem by 'programming' the camera to click the button itself, illustrating how their main priority has been to capture this moment of cosy family proximity as 'accurately' as possible.

Figure 6.5 was taken in the same way. For this German couple travelling without children, the bodily and aesthetic pleasures of cycling together in scenic surroundings were paramount. There are several images portraying their shared bike – a tandem – standing at rest. The picture is artfully choreographed to capture their actual movement through charming landscapes – nature as a 'place of mobility'. As the woman, who took virtually all the photos, says: 'I sat in the rear and all the time I had to look around and say "Stop! Stop!" I want that picture! [laughing] And then we stopped and sometimes I took my tripods out and, you know it takes me – whatever, say, 15 minutes? – to get it all set up and take the picture. And then we continued cycling' (Interview 10). Especially when one is sitting at the back, the tandem is a vision machine that constantly produces the elusive pictures she craves and spends a lot of time fixing on paper (see Larsen, 2001, for discussions of mobile sightseeing: the 'tourist glance'). The picture also exemplifies how every so often landscape images become hybrids of the 'romantic gaze' and the 'family gaze'. This particular scenery was certainly chosen to frame their shared movement in because it was considered scenic. And in this sense it shows how 'romantic' photographers picture their 'loved ones'.

These five photographs thus show how the 'family gaze' is put to work to portray idealized family relations and public places that appear natural and private. We have exposed the codified nature of the 'naturalness' of tourist photography. Photographs are highly staged and performed in accordance with a small number of scripts. The loving family is fashioned through eye contact, entwined bodies, playing bodies and solemn bodies. The attraction of such codes: 'is precisely this embrace of the conventions. Pictures that match such expectations give pleasure partly because their familial structures manage tension between the longed-for ideal and the realities of the lived experience' (Holland, 1991, p.4).

These photographs illustrate how performances are enacted to eradicate ambivalent memories. While celebrated for producing visions and memory, tourist photography's 'small world' of positive extraordinariness produces invisibility and forgetting. Tourist images produce 'calculated memory', the way one would like to be remembered and to remember places. They conceal as they reveal. They represent a reality that is a projection of their maker's desires. The perfect family and the perfect holiday may be a figment of the public imagination, but it has come to stand for something that ought to exist. Unhappiness and frictions are nowhere to be seen in people's rosy pictures of joyful moments and familial togetherness. People are keen to photograph for it is in the space of the photograph that they enjoy longed-for familial happiness. People gaze at the image and the imaginary family of their holiday gazes back.

Time and Memory

> [Making photos is] very important: no pictures, no memory almost. Some memories, but they fade very quickly; if you don't take pictures you forget, so if you take pictures, then you go 'ah, I remember this' and then you remember other things that you didn't take pictures of...(Interview 10).

The individual and collective memory work of reflection and recall tends to be organized around material objects (Radley, 1990, pp.57-58). Photography is one of many ways of achieving this, yet 'photographs belong to that class of objects formed specially to remember, rather than being objects around which remembrance accrues' (Edwards, 1999, p.222). The fascinating power of photography has always been bound up with a longing to fix the fleeting.

Fox Talbot sought to develop a technology that would cause nature's images to imprint themselves durably and remain fixed on paper. He described his invention of photography in 1840 as follows:

> The most transitory of things, a shadow, the proverbial emblem of all that that is fleeting and momentary, may be fettered by the spells of our *"natural magic"*, and may be fixed for ever in the position which it seemed only destined for a single instant to occupy...Such is the fact, that we may receive on paper the fleeting shadow, arrest it there and in the space of a single moment fix it there so firmly as to be no more capable of change (cited in Batchen, 1999, p.91).

Talbot describes photography's ability to fix nature's images, to arrest time and space on paper with unprecedented realism, as 'natural magic'. Photographic images appeared perfectly magical. The term 'natural magic' describes how modern science at the height of its rationality produces appearances that appear as wonderful *super*natural effects (Slater, 1995a; and see Crawshaw and Urry, 1997).

The enchanting 'natural magic' of photography was also desired as a means of capturing 'social worlds'. Ordinary people expressed the desire to have their 'loved ones' fixed forever in their own shadow. Elisabeth Barrett, wrote in 1889 to a friend in a letter that:

> It is not at all monstrous in me to say...that I would rather have such a [photographic] memorial of one I loved dearly than the noblest artist's work ever produced...it is not merely the likeness which is precious in such cases but the association and sense of *nearness* involved in the thing...the very fact of the *very shadow of the person lying there fixed forever* (cited in McQuire, 1998, p.14, our italics).

With the launching of the user-friendly, lightweight and cheap Brownie cameras in the late 1880s, *Kodak* was the major inventor of *tourist* photography as leisurely, mobile and light family-centred performance (see Larsen, 2003, chapter 3; West, 2000). With their extensive marketing campaigns, Kodak further scripted photography as a mechanism for the production of memories superior to that of human memory. As one typical ad said:

> The only holiday that lasts forever is the holiday with a Kodak...Few memories are so pleasant as the memories of your holiday. And yet, you allow those memories to slip away. How little you remember, even of your happiest times. Don't let this year's holiday be forgotten – take a Kodak and save your happiness...the little pictures will keep your holiday alive – they will carry you back again and again to sunshine and freedom (in Wells, 2001, p.144).

Kodak lectured the public, telling them that while few memories are as pleasant as those from holidays they only become proper memories when eternalised and materialized as photographic images. Otherwise they just 'slip away'. The camera keeps 'alive' fleeting moments of joy by transforming them into enduring images that transport the owner back 'to the sunshine and freedom' – again and again. Photography charms by providing 'imaginary travel' in embodied landscapes of memory.

We found that the fear of returning home from Bornholm without such memories of 'sunshine and freedom' meant that tourists regularly reached for their camera. They are anxious that they have not recorded family stories. Few express faith in unmediated memory, in their own mental and embodied 'pictures'. 'To have been there' is apparently no guarantee of remembering. No one said: 'Why take pictures of it, when I have seen it with my own eyes and sensed it with the whole of my body?'. The desire to arrest time and to capture memories is mentioned in almost all the interviews as the reason for the allure of photography.

A deep-felt desire to turn instantaneous and ephemeral tourism experiences, otherwise destined to exist only for a single moment, into eternal images, animates the camera work of tourists:

> I think it's an attempt to fix time, some experiences, a passage of time that has meant something to you, which you would like to hold on to as a memory (Interview 7).

> That, I think, is a very important dimension – stopping time. And the confirmation that you can sit five years afterwards and still hold on to what happened...That's profoundly satisfying; it's rather nice: you put time on hold (Interview 9).

The desire to stop time is the magical goal of tourist photography (Paster, 1996). The tourists long for the camera to immortalize their shared experiences for future pleasure, to arrest joyful moments of 'doing' such as playing on the beach, visiting a sight, eating at the summer cottage. The magic of tourist photography is the way it fulfils modern people's longing for immobility in an era of to use Bauman's telling metaphor, 'liquid modernity' (2000). The flux and flow of tourist experiences are: '(re)solidified', ripped out of time, into something that people can hold in their hands thanks to the magic work of the camera – a souvenir, a tangible memory, a form of control.

As Barthes shows so poignantly, personal photography works via sentimentality, through love and death; it is an 'order of loving' (2000). Whether the photos were stored in the processing envelopes, shoeboxes or organized in albums, people spoke of them with compassion and love. In all their funereal immobility, photographic memories are full of life, they keep alive. They not only fix, but also produce 'nearness'. People do not see a photograph of a 'loved one'; they only see the person. The photograph becomes a corporeal extension of the person pictured (Sontag, 1977; Paster, 1996, p.104; Barthes, 2000, p.7). When showing photographs, people say, 'This is my girlfriend and my kids', rather than 'This is a picture of my girlfriend and my kids'. Personal photos are predicated on this slippery relationship, or illusory co-presence of the represented and its referent. The photographs that touch us – with pleasure or grief – are transparent; the viewer is face to face with the person. This explains why there appears 'to be an almost insuperable desire to touch, even stroke, images. Again the viewer is brought into bodily contact with the trace of the remembered. We can say that the photograph has always existed, not merely as an image but in relation to the human body, tactile in experienced time' (Edwards, 1999, p.228).

Since photos are their irreplaceable family history, some people say that they are the first material possessions they would rescue in the event of a fire. An important part of them would burn if their photographs were destroyed by flames. Photos are extremely precious belongings, definitely not throwaway images, 'no matter how hideous the image may be' (Interview 14). As one woman said, to cut, tear or burn an image, is to cut, tear or burn a person once dear to one's heart. Discarding a photograph is an act of murderous hate (or indifference). Mavor describes it as 'a violent, frightening hysterical action, which leaves behind indexical wounds and irreparable scars' (1997, p.119). People dream, remember, hope, despair, mourn, gossip, love and hate with their photos. The space of photographs is inherently ambivalent because people respond to them emotionally. Barthes, writing on memory and photography, could not bear to reproduce the snapshot of his deceased mother. It was too personal and the reader would not understand his response to it (2000).

Personal photographs enable one to travel back in time, to connect with, and revive memories of, events and people through 'imaginary travel'. As material objects that halt time, photographs permit travel in past time. One man describes the fascination of browsing through the photo album as 'like going back in time, isn't it? Like a time archive, right?' (Interview 16). Tourist photographers are concerned with accumulating present personal experiences for *future* communal consumption. Their photographic practices of the now are about a changed tomorrow. This is especially evident in relation to the picturing of children, the main attraction of the 'family gaze'. The anticipation that their children and themselves will, at some point in the future, appreciate images of their former lives get their parent's cameras clicking.

The fascinating power of photographic images is not so much the pictures themselves, as the unseen stories that exist or can be activated beyond the frame in a future social context. As one interviewee says:

Well, it's the fact that you can get it out again. You evoke memories from the tour or experiences that you had in relation to that photograph, and it's fun to share them with the other participants; or with other people – you can talk about what you saw and what you experienced. It's the experience behind it [that is important]. Not so much the image itself (Interview 6).

While the interviewees expressed a remarkable faith in photographic memories, they were clearly aware of, and in fact valued, the complex fluidity and intertextuality of photographic remembering, of memory-travelling through holiday snaps. Photos are never in any simple sense containers of fixed memory-stories, stored there for good, waiting to be passively consumed. An active – and often collective – process makes photographic memories possible. Photographic images arrest life, but in people's actual *use* of them, whether silently or in conversation, they are enlivened and become full of time and life. Memory moves and lives in the body, and is much richer than images, but in order to be activated the 'dead accuracy' of photographs perform wonders (Pocock, 1982). This is not least the case in relation to the co-operative work of *sharing* memories of shared memories, which much tourism and photography are about. To quote one woman:

...A help to the memory...to trigger off something that you have in your mind, but don't really remember; and when you see the photograph, then HELLO! 'That's right, do you remember that and that and that...?' (Interview 7).

When the interviewees talked about photographic memory, it was often stressed that a photograph could set off a train of memories and memory-talk that moved far beyond what the image depicted. The image starts off the memory journey, but it is hardly the destination. Such work upon the image transcends the frame of the photograph and the visual. Many people said that pictures trigger off recollections about what happened before and after the event. The memories and meanings articulated through, and attached to, the moment fixed by the image, exhibit spatial and temporal flexibility. With the aid of the viewer's multi-sensuous memories and travel-talk, the single image can have many stories to tell. Some people stated that photographs invoke memories other than the visual ones such as the smell of landscape, the taste of the food, the temperature of the sea, the heat of the sun and so on.

Photographic memories and narratives travel in time as people move through life. As a father says, 'I'm sure that the pictures we have taken today will mean something completely different in, say, ten years' time, when the kids are teenagers and we have grey hair. No one knows what the future will bring' (Interview 5). While they are images of the past, photographs are always about today. 'Family photographs effect to show us our past, but what we do with them – how we use them – is really about today, not yesterday. These traces of our former lives are pressed into service in a never-ending process of making, remaking, making sense of, our selves – now' (Kuhn, 1995, p.16).

Many tourists stressed that the personal value of photographs increased with the passing of time. The charm of the family photo album, as a nostalgic medium, grows with age. As a woman says:

> ...I think it's good fun to take photographs, and I also think it's good fun looking at them. Also after a long while. In fact, even more so after some time. It's quite nice to look through the photos when you collect them. After six months or more you look at them again, and now they are really fun. In fact, photographs are like wine. The good ones become more valuable with the years (Interview 6).

As one interviewee said in relation to a discussion of images of children: 'the older you get, the more you want to see them' (Interview 6). Perhaps, the closer people move to their own death, the more intrigued they become with the immortalizing nature of the photo. Indeed the dark side of photography is loss, absence and death (Rose, 2003). Even while a joyful and sparkling life is celebrated, 'death' is always present, haunting family albums. Photography is a murderous act transforming living subjects into dead images of funereal immobility (Sontag 1977; Barthes, 2000). Photographs therefore provoke feelings of mortality in the spectator. When browsing through their celebratory album, people look right into the eyes of dead relatives, and they gaze up on their own once youthful bodies. The arresting of time by photography makes people acutely aware of the fragility of happiness and bodily ageing. The common melancholy experienced when one looks at one's former selves and (former!) 'loved ones', the jealousy of the photo's eternal youth, is what is so poignantly uttered in Wilde's *The Picture of Dorian Gray* (Wilde, 1951).

Since the reality of the photograph is a reality 'that has been' (Barthes, 2000), the charm of looking at photos is mixed with terror. This is particularly so with high mobility and the 'normal chaos of love', with friends and partners travelling in and out of lives at increasing speed. For many people, their photo collections surely represent lost contacts, broken dreams and bleeding hearts, as much as happiness. A lovely family photograph changes meaning when life turns into death and love is transformed into hate. Browsing through one's photo album can also be disturbing, because it portrays blissful and intimate life of which one has little memory. Even emotionally 'dysfunctional' families strive to exhibit tender family life, so problems are suppressed if only for the split second that the shutter is open (Holland, 1991, p.7; see Kuhn, 1995 for a disturbing account of this).

This may explain why almost all the people interviewed stressed that the primary audience for their holiday photographs was themselves. If they were displayed at all, it was in a casual fashion to very close family members and friends, but none of them talked about any organized showings or said that they looked forward to putting them on display. Some adults, particularly the 'younger' ones, found the whole idea of showing off their holiday pictures embarrassing: 'It isn't something that you put on show for your friends when they come around (laughing): "Look where we've been, have a look". I think, if we take photographs, it's mainly for, and because of, the children' (Interview 11). The two couples travelling together were the only ones to talk explicitly about the pleasures and traditions of showing photographs:

Tourist 1: It's also part of the holiday when we meet up one evening for a nice meal; when we have a very cosy and pleasant time seeing and talking about the four or five films we took.

Interviewer: Is it only you or do other family members or friends come around as well?

Tourist 1: No, it's only us that think it's fun.

Tourist 2: Of course, only us: our photos and us (Interview 2).

People are clearly aware that the snapshot world reduces experiences so much to private stereotypes that it is very difficult for other people than the participants to take an interest. As one father said rather frankly:

It can be incredibly boring, can't it? It's probably great fun for me to see my kids and myself, isn't it? But for my family and friends it's one of the most annoying things there is, isn't it? ...I remember my parents-in-law were in Thailand on a round trip lasting about twenty-three days...they brought twenty-five hours of video film home with them. All right, it was boiled down to eight hours. EIGHT HOURS – that's a bloody long time... I guess there were four hours of filming through the window of the bus. It was probably very interesting for them, right? But bloody hell... (Interview 19).

These photographers primarily produce and recount travel narratives to and for themselves. The spectators of tourist photography are the producers and actors themselves. Family-oriented holiday photos seem more akin to the diary than to the conventional public photograph. Like the diary, these photographs are not meant to be seen by anyone but those participating in the private event. There seems to be little appropriation of cultural capital through the 'family gaze'. Unlike postcards, they do not work through an 'envy economy'.

While most people have great affection for their photographs, they spend little time consuming them. According to research, they do so once a year or less (Slater, 1996, p.138). The interviewees say that they are always excited about how their films will come out, but that they very seldom look at their old holiday photos. Moreover, while people – especially women – express a desire to organize their photos in albums, in reality most families have piles of developed films 'lying about'. Why this somewhat peculiar emphasis on production rather than consumption? How can it be that people are fervent image-makers *and* affectionate about their photos, when they spend so little time 'consuming' them? While 'storying' has become characteristic of modern identity construction, and photography practices are ritually structured into various events, like tourism, that cannot be neglected photographically, the viewing of them is not: there are no social events that really call for the consumption of snapshots (Paster, 1996; Slater, 1995b, p.141). People take photographs at weddings, on birthdays and on holidays, but rarely look at photographs at weddings, on birthdays and on holidays. What is really important to people is not so much consuming as possessing them. One man explained how he hardly ever looked in his photo albums, but that walking past them every day was a comforting experience (Interview 7). The reassurance that one's images exist and are within reach seems crucial.

Slater suggests that the pin-board is increasingly taking over the privileged position of the photo album (in particularly among young adults), and several of the interviewed tourists explain that they adorn pin-boards, fridges and walls with their favourite photos (1995b, p.140). In such collages, holiday images are turned into everyday objects in domestic spaces and even in semi-public offices. Unlike the photo album, such displays reflect mutability and 'instantaneous time'. It is peoples' new friends, partner or wife, their latest baby or grandson, their latest holiday, that they display in this celebratory way. The pin-board exhibition is suited to short-lived relationships and life on the move, where people travel through places and are themselves 'travelled through' with increasing speed and frequency. As something 'erasable', 're-usable', something calculated not to hold on to anything forever, the pin-board shares it temporal economy with the videotape and the digital camera. They are suited to identity projects that revolve around the avoidance of fixation and keeping the story-line open what Bauman calls 'liquid love' (2003). Photography is becoming less covered by the thick dust of nostalgia and bound up with naturalizing the modern family.

Conclusion

It is often argued that tourism is essentially about having pleasurable experiences while one is away from home. Yet this study of tourist photography shows that it is fuelled as much by the desire to accumulate memories for future pleasure. Tourists are drawn to photography in order to produce memories. They anticipate that the camera will work magic transforming short-lived, fleeting experiences into durable artefacts that provide tickets to undying 'memory travel'. While they are a product of tourism and are caught up in the 'moment', holiday photos are not transient. Pictured tourism experiences have an enduring after-life. They become a vital part of life-stories and spaces of everyday life.

Holiday photography bridges the gap between home and away in another sense. Our quantitative and qualitative analyses of tourists' own photos showed that tourist photography is closely bound up with the staging of social relations and the transformation of places into private theatres of blissful family life. More than half the images collected portray friends and family members involved in intimate social life *and* attractions: beaches, landscapes, holiday homes, restaurants and so on. The 'ordinary' and the 'extraordinary', the 'private' and the 'public' intersect in complex ways in many 'family gaze' images. In fact, this gaze blurs these distinctions. The many 'family gaze' images exhibit intimacy and love more than anything else. They reflect people's desire and need to produce memory-stories that provide meaning, permanence and love in their social relations in an era of 'pure relationships'. To eradicate the 'normal chaos of love', the 'family gaze' naturalizes the intimate and loving family. Holiday images, for all their triviality, are precious possessions for people.

While family photos are inscribed with specific meanings, photographs are not receptacles of fixed memories, but objects around which *new* meanings are constantly produced here and now. The meanings of our photos are seldom static, because our life stories are characterized by flux and rupture as much as by stasis. It is the combination of photographic images and human work that produces memories that escape being nothing but photographic memories. They are not confined to the visible reality of the image: the single photograph triggers off the trip down memory lane; but memories that have little to do with the content of the image can be activated when the people involved work on them physically, imaginatively and through talking. When people engage with photographs they can become 'full of life'.

Chapter 7

Inhabiting, Navigating, Drifting

Up for an early morning bath, at the beach all day, bathing, building castles in the sand, collecting mussels at the beach, the children tumbling around in the sand, had lunch on the beach. Walked to our house, decorated the house with shells and stones, played cards with the children (Diary entry by a German woman).

Tourism and Mobility

We have discussed how tourist sites and places are staged and performed, especially through tourist practices. In this chapter we examine the role of tourist mobility in performing, producing and perceiving places. In doing this we conceive of tourist mobility as a 'performed art', with its own styles of relating to landscapes, sites and people encountered, perceived, experienced, made sense of and enjoyed (Adler, 1989a, 1989b). Tourist travels are the expressions of stylised 'modes of movement' employed as ways of sensing places.

The nature of corporeal mobility is an obvious characteristic of tourism yet often forgotten (Urry, 2002, p.152). If addressed, mobility has been treated as a pre-condition for performing tourism, a practical issue of 'getting there and around' rather than a topic in its own right (Pearce, 1988; Debbage, 1991; Fennel, 1996; Aronsson, 1997). In doing so, tourist mobility has been transformed into a black box, rather than a phenomenon *tout court*.

The fleeting, liquid nature of tourism has been held responsible for generating a superficial and shallow appreciation of places, cultures and people in tourism (Boorstin, 1962; Relph, 1976). Hence, tourism has been said to embody a desire for 'total mobilization' (Enzensberger, 1958); a desire for fleeing the 'horror of home' towards a destination set 'anywhere, anywhere as long as it is out of the world' (Baudelaire, quoted in Botton, 2002, p.32). Whereas it has been common to conceive of tourism as fuelled by desires for escape attempts, the embodied practice of these 'ways of escape' have not really been an issue (Rojek, 1993).

This chapter unpacks this black box of tourist mobility. It does so by asking 'why' and 'how' tourists move, rather than 'where'. Following the discussion in chapter 1 this chapter examines the spatialities produced through the discursive and embodied practices of corporeal movement. It treats movement as an ensemble of skilled practices involving reflexivity as well as various mobile technologies.

Deriving pleasure from movement has always been at the heart of 'the tourist experience', whether in the form of the sightseeing tour or in wandering, tramping and hiking (Adler, 1985, 1989b; Urry, 2000; Larsen, 2001; Solnit, 2001). Hence the spatio-temporal choreographies of tourism are: 'undertaken and executed with a primary concern for the meanings discovered, created and communicated as persons move through geographical space in stylistically specified ways' (Adler, 1989a, p.1368). Moreover, the different modes of movement are thoroughly inscribed with particular social and cultural codes. Writing about the history of sightseeing, Adler points out that even 'the traveller's body, as the literal vehicle of travel art, has been subject to historical construction and stylistic constraint. The very senses through which the traveller receives culturally valued experience have been moulded by differing degrees of cultivation and, indeed, discipline' (Adler, 1989b, p.8). Such discursive framings of how places and landscapes are (and should be) sensed as people move through them seem crucial to the pleasure derived from them. As de Botton observes, 'the pleasure we derive from a journey may be depending more on the mind-set we travel *with* than on the destination we travel *to*' (2002, p.242). Particular social and cultural 'mindsets' prefigure different styles of movement and the 'sensuous geographies' they imply.

Social and cultural studies of tourism have largely been conceived as a quest for spectacular places, other than home, and populated with exotic creatures, bodies and objects. They are sacred places or 'shrines' perceived to embody the Otherness of meaning (Hetherington, 1998, p.118). Mobility has been conceived through the point of arrival, the end of the pilgrimage. In this book we partly focus on family based holidaymaking in second-homes; a tourism that appears non-mobile. We then ask how 'laid-back' mobilities unfold and structure this. The intention is to grasp the 'travellings-in-dwelling' and 'dwellings-in-travelling' within the context of family based holidaymaking. We will see how the embodied movement of families inscribe particular spaces, landscapes and places with significance (Crouch, Aronsson and Wahlström, 2001).

This chapter brings out the embodied and hybridised mobilities of family tourists on holiday, as well as the discourses framing their perception of sites and landscapes. It does so through a close reading of diary accounts kept by families spending their summer vacation of one to three weeks in rented holiday houses on the Danish North Sea Coast. All these families are city dwellers from Northern Germany and the Netherlands. They plan and structure their holiday (mostly not in advance), and they use their car as the primary medium to 'get around'. The material comes from two sorts of diaries.

The first type consists of detailed registrations of the daily space-time-budget of 42 individual families throughout one week. The second type consists of more open descriptions of each of the days, together with a set of holiday snapshots from eleven German families (see list below). The outline diaries were handed to the families by the second-home agency when they collected their key to their house. They were informed how to carry out the writing of the diary and for what purpose it would be used.

In this chapter these diaries are read as accounts of what the families did (where, when and how they did what) as well as how they made sense of these activities. Especially the second type of diary allowed for 'thick description' and reflections upon the flow of the day. Often the diaries ended up being a patchwork of children's drawings, mental maps and diary accounts, thus turning the writing into a communal family project. In this respect the diaries also functioned as an artefact used to construct and re-produce the social bonds of the family. This also illustrates the place travel narratives occupy in re-producing family life.

Writing Tourism into Time and Space

While hegemonic views in social and cultural studies of tourism depict tourism as a quest for out-of-the-ordinary experiences, more 'mundane' and trivial sorts of tourist practices (such as mass package tourism or second-home vacationing) have often been excluded (Franklin and Crang, 2001). Hence the complex intersections between the realms of tourism and everyday life have not been much explored. Consequently the landscapes of tourism depicted by scholars have tended to oscillate between the figures of the traveller and the tourist, the cosmopolitans and the parochialists. Within this picture, family based second-home holidaymaking is viewed as a trivial counterpart to the otherwise serious business of doing tourism.

Löfgren similarly suggests that the characters of Phileas Fogg and Robinson Crusoe metaphorically encapsulate these contradictory aspects of modern tourism. The first epitomises the 'energetic and curious travellers who depart with the ambition to learn something, to widen their horizons'. The latter strive 'to create a utopian alternative to the humdrum of everyday life' (Löfgren, 1999, p.68). Whereas the flâneuristic Fogg represents a visual and disembodied experience of place, the Robinsonian quest is closely tied to the embodied use and practising of space.

What separates these two emblematic figures is how they relate to space. To the hyper mobile Fogg space is transformed into webs of extraordinary si(gh)t(e)s separated by the 'non-places' of the infrastructure (roads, railways, airports and so on). Hence s/he incarnates the commodification of space which Sack has described as a 'thinning out of space' (1997, p.9). Conversely, the Robinsonian tourist is a person who '"practises" the place, who uses it, experiences it, gives it social meaning' (Wearing and Wearing, 1996). The quest of the Robinsonian tourist is to create a space for 'laid-back' movement, a deliberate slowing down of everyday life. It is not an escape from home but for home. As shown in chapter 6 it is a quest for building a 'home away from home' (Jaakson, 1986; Williams and Kaltenborn, 1999), a utopian place where idealized version of extraordinary everyday lives and imagined family relations can be played out.

While such dualisms are persuasive and useful for identifying contrasting ways of sensing particular sites and attractions in tourism (see chapters 4 and 5) they can be problematic if they prevent us from understanding the multi-sensuous and hybridised movement. Places and landscapes are not encountered 'naked' but through the deployment of a variety of 'prosthetic' objects (Lury, 1997; Larsen, 2001; Parinello, 2001). Visual technologies such as cameras and VCRs are used not only to represent places and attractions, but also to choreograph and stage practices of family members and fellow travellers (see chapters 5 and 6). Mobile technologies such as bikes and cars are crucial not only to 'get around' but also to feel and discover landscapes and places (Larsen, 2001). Hence technologies are central to how people appear to grasp the world and make sense of it. They are decisive for how places are (and indeed can be) encountered and perceived. To move beyond such dualisms it is necessary to consider the manifold hybridities of, home-and-away, places and people, matter and mind, nature-culture, travellings and dwellings.

Inhabiting

Family based holidaymaking in (rented) second-homes is the major part of Danish tourism (as we saw in chapter 2). For urban families in Northern Europe, spending the holiday at the Danish North Sea Coast is one option among many. The holiday is seen as a way of exploring the region in which families have rented their second-home. 'The tranquillity and beauty of this place', 'the astonishing view of the troubled Sea', 'to explore the North', are common reasons given for choosing this kind of holiday (as in the opening sentences of the diaries).

As argued in chapter 6 family based vacationing is more concerned with the extraordinary ordinariness of personal social relations, than with documenting and gazing at spectacular sights. What matters is not seeing, gazing at or experiencing but rather *inhabiting* places. 'To build is already to dwell', Heidegger argued (2002, p.348). This Heideggerian link between building and dwelling is present in the accounts given in the diaries. One example is the frequent illustration by children that show houses, trees and sun-drenched lawns (Diary 4, see list of the diary-writers in end of this chapter). Another example is found in the diary account quoted at the beginning of this chapter. Here the decoration of the second-home house with objects of the beach (stones, mussels) unfolds as a significant task of the day (Diary 3).

According to Jakob von Uexküll 'every subject spins his relations to certain characters of the things around him, and weaves them into a firm web which carries his existence' (quoted in Ingold, 2000, p.174). By 'spinning relations' to landscape features, objects (and people), people transform environment into a home. Building castles in the sand and collecting mussels and stones for decorating the holiday house are not only child's play. They are constructive efforts to symbolically domesticate the stages of holidaymaking transforming the place of holiday into a home. 'My home is my sandcastle' as Löfgren epitomises Robinsonian tourism (1999, p.231). Hence places are valued not for their immanent qualities, but for their ability to serve as a safe haven for the family and as landscapes in which the family can inscribe itself and its social roles. Such tourism practices construct hybrid landscapes of home and away:

18/8
We arrived at [this place] where we have worked out that we were 12-15 years ago (...)
Here we found a lovely exclusive house, with small dunes of its own, just outside you
can view the sea. The first road of course took us to the beach. Off with the shoes, and
then off we went until we found a wonderful landscape of castles in the sand (Diary 1).

The joy of encountering the house and the anticipated views are what excites
the mother who writes this diary. The view of the house and the dunes immediately
prescribes what should be done, and what paths to be followed. The tracks of the
family's taskscape is already laid out in the landscape in advance as 'the first road
of course took us to the beach', where the family comes across not an astonishing
view, but a 'wonderful landscape of castles in the sand' – traces of already
anticipated holiday memories.

For most tourists the landscape of the Danish North Sea coast is well known
since they have already holidayed there in the past. It is a landscape invoking
emotions and associations with childhood memories, earlier vacations and so on.
The current vacation can be matched up against an accumulated body of memories
and desires of the eternal summers of an imagined past, as in this account of a day
on the beach:

21/8
Today beautiful weather. That means: Off we go, into the sea. For our daughter this is
the first time ever she will have a bathe in the sea. She is in Denmark for the first time,
and she is so enthusiastic about the dunes, sand and sea. We were here in DK ten times
already with our sons.
Pure sun – beautiful water – total relaxation (Diary 1).

The daughter (a latecomer who has not shared the ten preceding vacations here
with her parents and grown up brothers) is monitored and her reaction towards the
dunes, sand, sea and waves is recorded. Hence 'inhabiting' is an active and
constructive attempt to (re)inscribe members in the common family narrative of
eternal summers, 'pure sun – beautiful water – total relaxation'. The dense, almost
chemical formula, makes out the perfect recipe for the place transformed into an
imagined and fantasised home of eternal summers and happiness and the vacationer
into an inhabitant. This project of home making is also evident in how the relation
to the local population is represented:

22/8
When we had taken in supplies we went fishing. (...) We caught heaps of trout. The best
of course were the Danes. It's nice when you know where and how (Diary 1).

The local inhabitants in this account are valued because they can be
encountered as well known, and not some strange 'other'. They can be appreciated
precisely because 'it's nice when you know how and why' to get along with them.

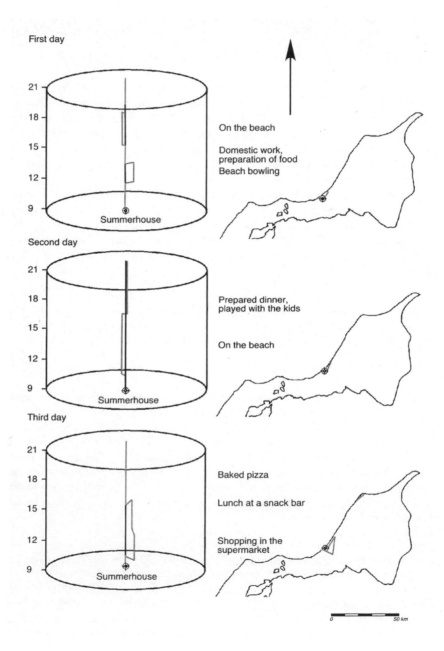

Figure 7.1 'Inhabiting' time-geographies

The family is not seeking spectacular sites of Otherness; they 'know how where and how'; they are not tourists; they are very much at home; they may even be more at home in the fantasy world of the holiday house than in their permanent daily residence.

This domestication of the holiday place, by these 'inhabiting' families is reflected in their daily spatio-temporal choreography. Figure 7.1 maps the trajectory of such a family. Virtually all days are structured around a repetitive pattern with the holiday house as hub. Time is spent together in the family, playing in the garden, preparing meals together alternating with visits to the nearby beach or eventually a crazy golf and a shopping trip for supplies to the nearest supermarket. The holiday is structured around activities in the immediate surroundings: the beach, house and lake. Each site constitutes the stage for a plurality of performances. Place is experienced as a 'thick place' (Williams and Kaltenborn, 1999; Sack, 1997), places to be inhabited and lived in. Hence sequential clock time is dissolved into a flow of lived time, a place to 'build up freed from civilisational constraints' (Diary 5), a place providing 'time for us and time to do things together with our children' (Diary 9).

The space surrounding the holiday house of the family is primarily explored through long walks, jogging and biking in the woods or along the beach, 'laid back' mobilities that enables the visitor to get familiar with and domesticate the holidaymaking scene. In some cases the 'good air', the wind, the view of the raging sea, the smells from the sea and the woods are mentioned. In other cases the exploration of landscape is imputed with a specific meaningful purpose: fishing, picking berries and mushrooms that can be prepared for the family meal later on. All activities unfold as part of the joint project of building a home for the imagined family.

This project of 'home-making' however falls short if the place does not afford a proper 'sense of home'. This family states:

18/8
In principle we like the house we have arrived at. Unfortunately it's located directly on the main road to [a provincial town]. We are quite unhappy with these surroundings and I am quite surprised as I have been lucky to have enjoyed nice vacations in [other coastal regions in Denmark].

We didn't find the track to the beach. So we took the car. But we were rather frightened of all the cars we found there! What an odd thing to do!!! After having seen the beach we went back to our house and tried to make ourselves at home, moved the beds, found out that the key doesn't fit the kitchen door. Must call the agency tomorrow (Diary 10).

The disappointment of this family is first and foremost the inability of the place to be a home. As the location of the house – and even the beach – turns into a hostile environment, the place loses its attraction. What we see here is precisely the limits of the project of inhabiting: the symbolic significance of sites, landscapes and places inhabited are embedded in the particular narratives of the individual families and couples, their beings and doings, their project of home making. If the place resists and does not allow for the construction of a imagined home - this laid back 'mode of movement' falls short and the place loses its meaning.

Inhabiting is however not the only way of relating to landscape and place. In what follows, two other 'modes of movement' are explored.

Navigating

While inhabiting gains its meaning through an annihilation of time through space, the picture changes when we consider how sightseeing trips are organised and planned. Rather than 'inhabiting' places, the practices related to sightseeing are defined through the point of arrival, the si(gh)t(e) visited. Like the pilgrim the nature of a sightseeing trip is 'getting there', and the trick is to find one's way, to *navigate* space. One family describes their first car trip around the region:

> 23/8
> The weather is not so good today that we necessarily have to stay on the beach. So this is a good opportunity for a sightseeing tour.
>   After a shopping trip in Blokhus, we visited Blokhus-Candles where we had to fill up deposits with homemade candle-lights. Then we drove on the beach to Løkken...and went further - to Lønstrup. The visit to the lighthouse here and the cliffs around it is a "must" for us. It is a good place for the children to run and climb the dunes, and Wolfgang documented the changes of the coastline since our last visit with photos and video camera (Diary 9).

While the plotting of this car trip onto a map would show a trajectory following the flows of most tourists in the region, the trip derives its meaning through the 'collection' of particular places passed en route, in the form of souvenirs, photographs and vistas. Often the trips are thematised (such as 'beaches', 'birds', 'collection of mussels', 'handicrafts'). Hence, they require serious planning and calculation of time. Navigating is involves collection, exploration and documentation of the places passed. This shapes the style of writing in the diaries, as the descriptions of the practice of navigating implies that the trip can be accounted for in points shortly summarising the sites visited and the reasons for going there, as the following indicates:

> 23/8
> We drove south-west: Slettestrand, Fjerritslev, Torup Strand and Bulbjerg [beaches and villages along the coastline]. If anything Slettestrand is boring. We inspected Han Herred Nature Center [Hands-on natural history museum], and found it very good and informative. Then we went further to Torup Strand, where we lucky to see the arrival of fishing vessels, and how they were pulled upon the shore. Also here we bought fresh fish for our evening meal (fortunately we have a built-in refrigerator in our car). Finally we drove to the bird cliffs at Bulbjerg which we inspected from the upper side (among other things we spotted a gullery) (Diary 7).

Plotting their route almost in the form of a travel guide, this diary displays that they visit sites to 'inspect them', gain 'good and informative' knowledge, 'spot' the local wildlife. The car is not merely a means of transport, but a mobile machine extending the capability of the family to track their way into unknown territory and encounter or strange places. Through the car the family is transformed into an 'expedition group', finding their way, plotting the right course, navigating through space, and classifying the sites encountered. To navigate means to departure from the places of the well known and in set course towards points and places at which arrival is anticipated. It is as this objective is realised that the movement gains its meaning. Hence, navigation requires a rigorous organisation of time and space, in which places to visit are planned in advance, possible routes considered, times of arrival scheduled. It involves a multiplicity of objects such as maps, clocks, and guidebooks, to provide images and knowledge against which the knowledge gained en route can be matched up. Movement is experienced on the map as well as through the landscape, and possible routes are being revised and reshaped when planned routes fail, as this account of a failed sightseeing trip shows:

30/8
Trip to the Skagen and the Danish Baltic Sea coast.
   Because our guidebook said that the fishery auction would be a beautiful experience, we drove off very early (at 5.00). However that thing with the auction was a mistake. In the corner of the empty hall between a couple of cases, 2-3 people ran around. Despite this, we got a fish for display. As we were here already we then went off for "Grenen", where the waters of the North and the Baltic Sea "clashes". It was a beautiful morning trip, and we walked for a long time through the dunes. When we finally got back to our car, we saw the first bus in the morning arrive! Thank god that we arrived this early. On the way back we followed the Baltic Sea coastline to Limfjorden. We are very glad that we decided to rent a house on the North Sea Coast as the beaches here were rather boring, and also looked like a bad place to bathe from. However we did collect a lot of mussels etc. at the beach (Diary 7).

Setting out the anticipation of 'a beautiful experience', this clearly fails because of the guidebook's misinformation. However, plans for the trip are quickly revised and the family set off for the main tourist sight in the region and arrive before other visitors. Failure is transformed into success as '...we saw the first bus in the morning arrive' and 'we did collect a lot of mussels'. The central cultural codes running through these accounts are framed along the same lines as the 'Grand Tour'. Gazing at places is educational. Places are appreciated for their immanent qualities as sources of edification. To see places is to learn from them, explore them and value them for what they are (or fail to be).

Throughout the accounts on navigating, we see that mobility is closely tied to the points in space that the path cuts through. The trip gains its meaning and sense through these linked points. The places visited are evaluated and judged according to how beautiful, boring or instructive they are. Depending on the technological capabilities and planning skills of the family, the geographical trajectory can have a wide spatial horizon.

Figure 7.2 maps the trajectory of a family with two teenage daughters navigating their way through their holiday. The trajectory also reflects that their explicit purpose of is to go 'to explore the North' as they put it, and their choice of second-home residence is from the start informed by the anticipation of exploring Sweden and Norway. As such the 'time-geography' depicted in Figure 7.2 is the rational outcome of an attempt to plan and schedule as many different visits as possible. Eventually their navigation collapses as they get lost on their way between two sightseeing trips. The brief comments in the diary by one teenage daughter hint at an emerging family crisis:

Day three
    10-11:   Already in the car. On the way to Hjørring [provincial town].
    11-12:   Shopping in Hjørring (apparently there's some kind of party or festival
                there).
    12-13:   By car to Lønstrup.
                Visiting the church [Mårup], having lunch in the rain.
    13-14:   By car to Rubjerg Knude [cliffs at the edge of the sea] nearby.
    14-15:   Walking, *surviving*, sand all over the place.
    15-16:   In the car – driving.
    16-17:   Drinking tea in the car, counting German and Danish cars (506).
                Location? Somewhere far away (not Rubjerg Knude).

The sardonic tone also exhibits how navigation has to be embedded in the narrative of the family and the project of re-producing its social bonds. This may also explain why the accounts in the diary are filled with observations and monitoring of the children. Observations that are not merely a question of joy in the playing children, but rather a disciplining and regulating gaze on the children and their appreciation of the sites visited, or an anxiety that disengaged children might spoil the navigating. Hence, a common conclusion in the diaries about the days on the move goes as follows: '…and the children liked it too' (Diaries 7, 13).

Drifting

Finally, there is drifting. This refers to the pleasure of movement itself such as glancing at the passing landscape, sensing the kinaesthetics of corporeal movement, or simply heading out into the unknown (Larsen, 2001). Experiencing places through walking, climbing or running, historically developed its own cultural and social codes emphasizing effort and slowness (see Adler, 1985; Urry, 2000, p.55ff; Macnaghten and Urry, 2001).

At the heart of this is an opposition to the disembodied visual experience of tourist sightseeing or the 'navigation' between sights discussed above. Vision may be an important yet not exclusive way of sensing places. The sounds, odours and touch constitute particular 'senses of place' (Tuan, 1976). They contribute to particular geographies of tourist places and are constitutive to the kind of 'mindset'

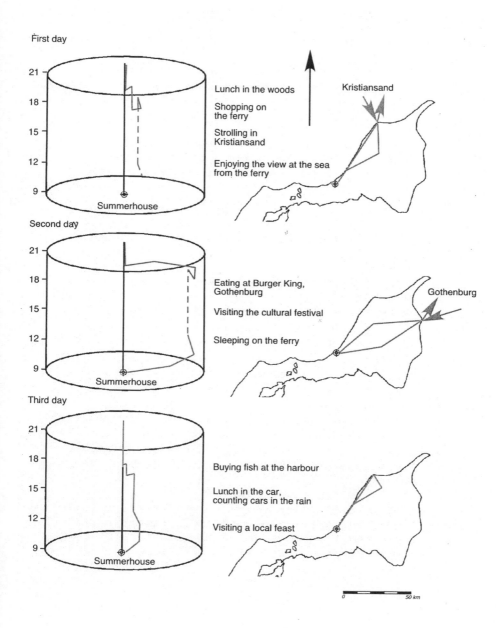

Figure 7.2 'Navigating' time-geographies

that prefigures some forms of hiking, biking and car driving. Within the context of second-home tourism this 'mode of movement' appears as momentary incidents of intense joy in the movement itself, of sensing the wind and the rain touching the skin, of the resistance offered by the dunes walked through or the hills conquered by bike.

This pleasure of drifting is present in accounts dominated by the logic of inhabiting and navigating. This can be illustrated by this account of a sight visit from a young couple:

> 27/8
> Today was an extreme day. Before breakfast we went down to the beach, and watched the big waves in the raging sea. After breakfast we drove to Lønstrup and the village nearby, at the edge of the cliffs, because we liked that place so very much. We had anticipated that we would run along the coastline to Rubjerg Knude [a big dune a couple of miles south further south]. However we had not thought that the strong storm would throw so much sand up into the air. The drifting sand was so tremendous that we had to break off our walk after a short while. Nevertheless we went to Rubjerg Knude to have a short look, but the drifting sand was so intense that we could not even see the lighthouse (Diary 5).

The sight is visited primarily because it is 'the village nearby' and 'we liked that place so very much'. The purpose of going there is to go for a run, though it is recorded that the drifting sand was so intense that 'we could not even see the lighthouse'. The purpose of visiting the church and the lighthouse is not to see this particular well-known tourist sight, but to 'run along the coastline'. Hence, the mode of experiencing this particular landscape through the moving body, and not by gazing like everyone else, is the very point in being there to this couple. The mode of movement is not incidental, but a specific style of experiencing this particular sight in opposition to the prescriptions of the guidebook.

Also motorised mobilities produce moments of pleasure. 'The thrill of racing in the car along the beach' (Diary 8), or 'enjoying the view and the wind from the deck of the ferry' suggest the same kinds of pleasures as 'running along the coastline'. Particular sites may even presuppose particular modes of movement to be properly enjoyed:

> Slept until 12. Got out of bed. Tried to find our way to the beach on our bikes. The sun was shining. Muttering over the cars, trying to figure out how many there were.*
> Back to the house.
> Driving down the beach in the car ourselves (see!) * ☺. (Diary 5).

Car driving may also include the pleasures of drifting. To this couple the beach is not experienced as a joyful site until they have 'learned' to encounter it in the car (note the 'smiley' used to indicate this solution in the quotation). The sights, sounds, smells and bumps of the car are transmitted to the driver together with added soundscapes (radio, music), the comfort of the seat and convenience of the heating/cooling system. In so doing, the car itself becomes a central artefact for experiencing places, landscapes and sights (Bull, 2001). As one family navigating along the coast writes, 'the best part of the trip were the views of the beautiful landscapes as we drove down the coastline' (Diary 2). Hence, drifting produces pleasures that escape both the logic of inhabiting and navigating and follows its own logic and prescriptions, the freedom and pleasure of movement along the 'open road' (some of this will be further examined in Featherstone, Thrift, Urry, 2004).

'Laid-back' Mobilities

We have discriminated three different styles or 'modes of movement', each rooted in particular social and cultural narratives that inform how tourists inhabit, navigate, and drift through space.

According to Löfgren, the 'vacation is an area in which fantasy has become a major social practice' (1999, p.7). As shown the various styles of mobility prefigure how people sense and make sense of places and sites, as well as how these place encounters are framed within divergent social and cultural codes. The travel diaries and the accounts of the particular 'modes of movement' they contain, show how the practice of movement is rooted within the project of the family and the attempt to (re)produce the family unit while on holiday. While navigation is based on a binary symbolic dichotomy between home and not-home (to be explored and encountered), people inhabit places by 'spinning' such relations to objects that signify an imagined and utopian home.

The three different modes of movement are fundamental to how landscapes, places and sites are sensed. Hence, corporeal movement is performed through social and cultural discourses imputing also stylistic constraints to the way that movement is performed. The enactment of these 'modes of movement' prefigures not only how places are sensed and perceived but also in the shaping and sensing of movement itself.

Just as the tourist sites of the navigating tourists are appropriated through a 'spectatorial gaze' the holiday world of the 'inhabiting' families are seen through a 'family gaze' (see chapters 5 and 6). 'Drifting' is fundamentally in contrast to both the disembodied visuality of 'navigating' and the parochialism of 'inhabiting', glancing at the moving landscapes while in motion or sensing them 'on the move'.

Inhabiting, navigating and drifting are practices that produce different tourist places. Mobility is not tantamount to spatial movement from one point to another. Tourists' movement in space is not incidental but are ways of encountering landscapes and places through the deployment of various styles of movement. Tourists navigate to find their way to the heritage sight, the petrol station or hamburger restaurant. They inhabit the beach, the car and the cottage lawn and fill these places with social life and meaning. They drift absent-mindedly, though open to the multi-sensuous impressions derived from the passing landscape or townscape. It is through the employment of such 'laid-back mobilities' that sites, landscapes and places are produced, performed and perceived by tourists.

List of Diary Writers on the Northern Coast of Jutland

(1)   Family consisting of a mother (age 52), father (age 54), son (age 26) and daughter age 12.

(2)   Family consisting of a mother (age 39), father (age 44), daughter (age 18).

(3)   Family consisting of a mother (age 38), father (age 40), son (age 11), son (age 9).

(4)   Family consisting of a father and mother (in their 30s) and three children under 10.

(5)   Couple consisting of man (age 33) and a woman (age 31).

(6)   Couple consisting of a man (age 26) and a woman (age 26).

(7)   Family consisting of a father (age 39), daughter (age 15), daughter (age 13), son (age 10).

(8)   Family consisting of a mother (age 38), father (age 39), daughter (age 6) and son (age 9).

(9)   Family consisting of a mother (age 34), father (age 37), son (age 3) and daughter (age 5).

(10)   Family consisting of a mother (age 43), father (age 39), and daughter (age 14).

(11)   Family consisting of a mother (age 40), father (age 40), sister (age 32), grandmother (age 64), daughter (age 14), daughter (age 11), daughter (age 7).

Chapter 8

Places, Performances and People

Every voyage is the unfolding of a poetic. The departure, the cross-over, the fall, the wandering, the discovery, the return, the transformation (Trinh T. Minh-Ha, 1994, p.18).

The 'New Mobility'

This book is about the making and the consuming of places in the contemporary world. Although the case-studies happen to be located within Denmark, the general arguments about the intersections of places, performances and peoples are much wider than this. It is shown that those who are visitors to a place in part produce that place through their performances. Places are not seen as authentic entities with clear boundaries that are just there waiting to be visited. Places are intertwined with people through various systems that generate and reproduce performances in and of that place. These systems comprise networks of 'hosts, guests, buildings, objects and machines' that contingently realise particular performances of specific places.

The studies reported here further develop what we might describe as the 'new mobility' paradigm that is developing within the social sciences. Recent general contributions include Urry (2000), Cresswell (2001), Riles (2001), Verstraete and Cresswell (2002), Amin and Thrift (2002) and Degen and Hetherington (2001). Examples of new mobility analyses of various leisure and tourist practices include Solnit (2001), Macnaghten and Urry (2001), Pascoe (2001), Coleman and Crang (2002b), Sheller (2003), Crouch and Lübbren (2003) and Smith and Duffy (2003).

This new paradigm rejects three characteristic positions. First, there is the rejection of the sedentarist metaphysics of humanist geography that locates bounded and authentic places as the root of human identity and experience (Cresswell, 2002, pp.12-15). Also criticised by the new mobility paradigm is the nomadic metaphysics that celebrates metaphors of mobility and flight, seeing mobilities as progressively moving beyond both disciplinary boundaries and geographical borders (Cresswell, 2002, pp.15-18; Urry, 2000, chapter 2). And third, the 'cultural critique of placelessness' associated with Augé is also found wanting (1995). This position inappropriately treats places of mobility as without significance and meaning, as spaces only to pass quickly through but which are not practised or performed or stabilised.

The 'new mobility' seeks to move beyond these particular metaphysical positions, to view places as significant to those living in them but also to those who visit. It sees places as contingently stabilised sources of deeply held meanings and attachments but where these stem from networks that enable particular embodied and material performances to occur, performances normally involving both 'hosts' and 'guests' (see Smith, 1989). So places are not fixed and authentic, nor do nomads overwhelm them, nor do places of movement evade practices of place stabilisation and significance.

The new mobility paradigm has sought to reveal this 'contingent mobility' through detailed social and geographical analyses. These analyses have been inflected by the cultural and spatial turns in the social sciences, and by recent analyses of the body, of performance and of objects. In this book we have sought to direct this new mobility position towards tourist practices, seeing these as material, embodied, contingent, networked and performed.

Moreover, in such performances there is no simple and unmediated relationship of subject and object, presence and absence. There is what has been termed a hauntingness of place, through voices, memories, gestures and narratives that can inhabit a place for both locals and for those visiting (see Degen and Hetherington, 2001). These ghostly presences of place are in-between subject and object, presence and absence. These presences can be described as the atmosphere of place, such as Allinge, Hammershus or Skagen, that is irreducible either to their physical or material infrastructures or to the discourses of representation.

In the next section we examine the shift in the nature of place as they have very broadly moved over time from 'land' to 'landscape'. We note some consequences of this shift and of resistances to it within Danish tourism. The metaphor of the sandcastle is then readdressed since this effectively captures our main line of argument throughout this book. Subsequently we examine two themes that this metaphor reveals, of the need to research the proximities involved in movement, and the methods necessary to research people, objects and images that are 'on the move'. The book ends with an account of what we take to be the main lessons of these studies for the 'new mobility' paradigm.

Touring Places

Wordsworth amongst others distinguished between the places of land and of landscape (Milton, 1993; Buzard, 1993). In the former place is a physical, tangible resource that is ploughed, sown, grazed and built upon. *Land* is a place of work conceived functionally. To dwell on land is to participate in a pattern of life where productive and unproductive activities resonate with each other and with particular tracts of land whose history and geography will be known in detail. There is a lack of distance between people and things. Land is 'ready-to-hand' (see Ingold, 2000).

By contrast place as *landscape* entails an intangible resource whose definitive feature is appearance or look (Milton, 1993). Landscape emphasises leisure, relaxation and the visual consumption of place especially by those who are 'touring'. Areas of wild, barren nature, once sources of terror and fear, transform into landscape, what Williams terms 'scenery, landscape, image, fresh air', places waiting at a distance for consumption (1972, p.160; Macnaghten and Urry, 1998). From the eighteenth century onwards a specialised *visual* sense developed in western Europe, based upon various technologies, the camera obscura, the claude glass, guidebooks, the widespread knowledge of routes, the art of sketching, the balcony, photography and so on (Adler, 1989a, 1989b; Ousby, 1990).

Wordsworth noted by 1844 that the very notion of landscape was a recent development in England. But within a few years of this even houses were being built with regard to their 'prospects' as though they were a kind of 'camera' (Abercrombie and Longhurst, 1998, p.79). The language of views prescribed a particular structure to the experience of place, as land gave way to landscape also in France as well as England (Green, 1990, p.88). According to Buzard, Wordsworth's poem *The Brother* thus: 'signifies the beginning of modernity...a time when one stops belonging to a culture and can only tour it' (1993, p.27).

The quintessential location of place as landscape occurred in the English Lake District from the late eighteenth century onwards. A place of 'land', of inhospitable terror according to Daniel Defoe, transformed into 'landscape', a place of beauty, emotion and desire (Urry, 1995). This place only became part of England when at the end of the eighteenth century onwards many visitors, especially artists and writers, travelled to it from the metropolitan centre. These visitors, with their poetic reassessment in terms of the picturesque and the sublime, brought the Lakes closer to the centre of England. Land turned into landscape through artists and writers 'moving' the Lake District *into* English culture. The Wordsworths, Southey, Coleridge and so on became celebrities in an area previously without national celebrities. They were major tourist attractions especially for metropolitan visitors who repositioned the peripheral area of inhospitable land, closer to the centre, almost part of a metropolitan nature, of a much desired landscape.

Likewise the Alps before the end of the eighteenth century had been regarded as mountains of immense inhospitality, ugliness and terror. But they too became 'civilised' as landscape; Ring describes how the Alps are not simply a place of land. They are rather a 'unique visual, cultural, geological and natural phenomenon, indissolubly wed to European history'; they are part of that history as they shift from land to landscape (Ring, 2000, p.9).

We have seen how places of land in Denmark, especially Bornholm and Skagen, also became places of landscape. In chapter 5 we noted how Bornholm's 'wild' and 'rough' landscapes and seascapes had been little praised before the arrival of tourists. In the late 1790s, Bornholm was described as the most horrible and frightful island on this entire earth. But soon this inhospitable place of land turned into a place of landscape, as by the mid nineteenth century city-dwelling painters were flocking to the island from all over Europe and moved it closer to the metropolitan centre.

Likewise we saw in chapter 4 how in the 1860s an artisans colony was formed in Skagen. The paintings of the sea, the shore and its people were transported into the public domain through exhibitions, writings in newspapers, journals and guidebooks (Lübbren, 2001, 2003). These representations turned the place of the beach as land into landscape. Artists scripted and choreographed the beach as a stage to gaze at, as landscape to be visited and to be enjoyed by those touring.

We also saw in chapter 5 how the technology of photography played a seminal role in shifting from land to landscape. Touring places and photography commenced in the 'west' from around 1840 with Louis Daguerre and Fox Talbot more or less simultaneously announcing the 'invention' of the camera (Schwartz and Ryan, 2003). In 1841 Thomas Cook organised what is now regarded as the first packaged 'tour' using the new modern technology of the railway (Lash and Urry, 1994, p.261). The 1840s then is when the 'tourist gaze' emerges, combining the means of collective travel, the railway; the desire for travel, through connoisseurship of different landscapes; the techniques of visual reproduction, photography; and the organisation of travel, through the organisational innovation of Thomas Cook (see Lenman's account of how various British places were remade through landscape photography; 2003; Osborne, 2000). Chapter 4 has shown how photography and new forms of transport were central to the remaking of northern Jutland as a place to be toured from the later nineteenth century onwards.

Moreover, as visual consumption becomes increasingly central to the constitution of place as landscape, so this takes on an abstracted, disembodied quality. E.M. Forster in *Howard's End* characterises the process by which certain places become nomadic or cosmopolitan: 'Under cosmopolitanism...Trees and meadows and mountains will only be a spectacle' (E.M. Forster, 1931, p.243; Szerszynski and Urry, 2002). Places become cosmopolitan, nomadic spectacles. The shift to a *visual* economy, that nature and place are to be looked at rather than used and appropriated, 'de-substantialises' place, making if ever more difficult to trace values back to their place of origin (Smith and Duffy, 2003, p.162). Each locality then becomes not a unique place, with its own associations and meanings for those dwelling or visiting, but a combination of abstract characteristics marking it out as similar or different, as more or less scenic than other places.

The modern subject is a connoisseur of place, as nature is transformed into landscape, comprised of images of trees, meadows and mountains known about, compared, evaluated, possessed, but not places to be 'dwelt-within' as land. The language of landscape is then a language of mobility, of abstract characteristics stemming from a heightened aesthetic reflexivity (Lash and Urry, 1994, chapter 3; Crouch and Lübbren, 2003, p.10). It is not just that such mobility is necessary to develop the capacity to be reflexive about places as landscape. It is also that landscape talk is itself an expression of the life-world of mobile groups including many tourists and environmentalists (Szerszynski and Urry, 2002; Mowforth and Munt, 2003). Thus almost all places across the world are 'toured' and the pleasures of place derive from the emotions involved in visual connoisseurship of different places. This involves the emotion of movement, of bodies, images, and information, moving over, under and across the globe and reflexively monitoring places through their abstract characteristics. Those mobilities, a 'fluid modernity' according to Bauman, have produced a widespread capacity for the judgment of landscape (2000). This is a judgment from afar, possessive and abstract involving what Chaney terms not the prolonged gaze but multiple glances (2000; and see Adler, 1989a, 1989b). This book has interrogated the shift from land to landscape and its consequences for various places across Denmark. We examined contrasting examples of landscape as many areas in Denmark came to be toured and subjected to connoisseurship from afar.

However, we have also seen that Denmark, like various other Nordic countries, has produced a rather distinct version of this shift. There developed a way of practising tourism 'home from home'. There is the recreation while apparently 'away' the dwellingness of family and domestic life. Such holiday home tourism is a resistance to place as landscape, a way of preventing the practices of touring from fully moving into landscape mode. We have seen in Denmark that many tourist places are familiar, ordinary, slowed-down places and practices of dwelling. They reconfigure home and away, occasionally at least producing what Thrift terms: 'an expanded awareness of present time' (2001, p.35). We noted that ten per cent of Danish households own a holiday home and many more have access to them. This is a highly state regulated form of Danish travel. No foreigners may buy such houses, there are restrictions on how many can be owned by Danes, and there are restrictions of building as well as legislation to prevent holiday homes being turned into businesses.

But this state form enables a tourism in which, as we have seen, production and consumption occur in close proximity. The holiday home constitutes a site for 'enthusiasms'. Using the technologies of cottage, car and camera, the family produces and consumes its own products of family life, being home while away (see Hoggett and Bishop, 1986, on 'enthusiasms'). One particularly popular site for such coterminous production and consumption is in the family, collectively building and consuming a sandcastle on the beach near the holiday cottage. There is something about the hauntings of a deeply familiar place that can draw families back time and time again to that same holiday home (Degen and Hetherington, 2001). Good memories can haunt such a place, a still life in nearly present time (Thrift, 2001).

Sandcastles

In this book we organised our analyses of place and performance through the metaphor of the sandcastle (see especially chapter 1). Sandcastles, we have seen, are tangible yet fragile constructions. They only come into existence through drawing together certain objects, mobilities and proximities. Its building involves children and parents working together, placing sand, buckets of water and objects for decoration and stability within a network. The resulting collective family performance is a castle of sand towering over the beach and drawing together memory flows, objects and matter. The humble sandcastle results from and symbolises a collective family performance within the distinct space of the pleasure beach (see chapter 4 on the beach as classical tourist trope).

This 'place' of the sandcastle contingently stems from various intersecting mobilities. There are imaginative mobilities in which during long winter nights people dream of sun-drenched summer beaches, the globally universal place to play in the contemporary world. There are corporeal mobilities, such as the journey to a holiday region, a day trip, and the dense choreography of a family moving around and building the sandcastle. Then there are mobile objects such as fish, stones and mussels at the shore or on the beach that may have travelled thousands of miles, waiting for their starring role. The tools for building, such as buckets and spades, are brought in the family car (maybe from Germany) but will have been manufactured and transported from China or a similar low-wage country.

More broadly, performing the sandcastle necessitates the road network, widespread access to private cars, holiday housing, camping sites, beach hotels, restaurant/ bar districts, marketing, internet, advertisements, public holiday acts, planning legislation, road maps, guidebooks, ideologies of domesticity, the nuclear family and so on. These are all elements of the networks that stabilise and regulate the sedimentated practices that result in the humble and transient sandcastle on the beach. This transformation of the beach into a social and performative space is unthinkable without these multiple networks stabilising and reproducing it.

These multiple mobilities moreover necessitate proximities. The making of a sandcastle is a social project involving face-to-face, body-to-body proximity among the family members who both construct it *and* act as an impressed audience. The family is simultaneously producer and consumer of the sandcastle. For a couple of hours the castle in the sand forms centre stage for the performance of play and for the applause of an admiring audience. The sandcastle, the sea and the sun are central to staging a carefree 'timeless', not to be forgotten, family afternoon. Such an afternoon on the beach involves assembling: 'the distant in time and space. [The beach] was a box in the world theatre' (as Benjamin analogously said of the nineteenth century drawing room: 1973, pp.167-8).

These elements provide the backcloth for moments of pleasure to be remembered and recorded, before the tide of history washes the castle away. Through the sandcastle, the space and materiality of the beach is domesticated, occupied, inhabited, embodied. As noted in chapter 1, the sandcastle transforms the endless mass of white, golden, fine grained or gravelled sand into a habitat; a kingdom imbued with dreams, hopes and pride. Nature is reconstructed as the social space of an embodied family performance. As Thrift writes more generally: 'The body produces spaces and times through the things of nature which, in turn, inhabit the body through that production' (2001, p.47).

But the sea rises and slowly the fortifications erode and the family leaves. Waves role gently on the shore removing all trace of the day's performance. All is washed away and the castle remains only in the memory of the family, in the photographs, and maybe in anticipating the next day back on the beach. Only the photographs remain to capture that moment in the sand.

The sandcastle captures many of the tourist performances that we examine in this book. The sandcastle is a metaphor for tourist places generally. We have seen that such places combine together five components: physical environment, embodiment, sociality, memory and image. The existence of a particular physical environment does not itself produce a tourist place; they are nothing but potential, dreams of something that may happen. And this is just like a sandcastle. A pile of appropriately textured sand is nothing until there is embodied activity, sociality especially around family life, memory especially as recorded photographically, and image of places. Indeed places only emerge as 'tourist places' when they are appropriated, used and made part of the memories, narratives and images of people engaged in embodied social practices. Such places are, as we have seen, inscribed in circles of anticipation, performance and remembrance.

In particular the beach is both central to this metaphor of the sandcastle and to several of the tourist places examined in this book. We develop an ethnography of the beach as a distinct kind of place, identifying those performances that enact, transform and play up against various discursive blueprints. The beach is a place of diverse embodied practices, including those of family life involved in symbolically making itself (often out of various fragments) as the sandcastle gets made (see chapter 4). We examine the nature of the family gaze focused around the performance and memories of family life upon the beach and especially in and around the holiday home (see chapters 4 and 5). Photography captures those moments that soon will wash away, to hold onto memories that easily escape (see chapters 5 and 6; and see Chambers 2003, on family photograph albums). 'Holidays last longer in snapshots' (Kodak ad, cited in Schroeder, 2002, p.74).

Places entail multiple mobilities, what we call inhabiting, navigating and drifting through various time-spaces (see chapters 3 and 7). The mobilities involved in tourist places are like those in the case of sandcastles, both very limited and very wide ranging. We show how complex networks have to be formed with 'consumers' in order that tourist places happen; and indeed how such networks can often fail to produce appropriate experiences because networks are not localised around an apparently given and bounded 'destination' (see chapter 2).

A sandcastle is not fixed and given but is fluid and changing. Tunnels and towers may collapse as the sun shines; the rising tide may cause water to penetrate the ramparts surrounding the moat; the fortifications may get undermined. The work of erosion and sedimentation may slowly alter the sandcastle or there may be sudden ruptures as the walls collapse. Erosion and sedimentation are also to be found in tourist places. We show how new sedimented practices can overlay existing practices, as with the transformation of Skagen from artist's colony into tourist site (see chapter 4), that places can get eroded such as some holiday home tourism (see chapter 2), and that new tourist places are always just around the corner as with Viking tourism developing in Roskilde (see chapter 2). Sandcastles can turn out to be transient castles in the sand.

We thus view places as economically, politically and culturally produced through the multiple networked mobilities of capital, persons, objects, signs and information. And it is out of these complex movements that certain places to play, of holiday homes, beaches, harbour areas and so on, are contingently assembled. They are 'produced' not only out of the enterprises and organisations that happen to be 'present' within a given locality. The realm of relevant enterprises and organisations is far wider than this. So places are not fixed or given or simply bounded. It is more profitable to see them as 'in play' in relation to multiple mobilities and varied performances stretching in, through, over and under any apparently distinct locality (as we showed in chapter 2 with regard to how business networks in tourism are non-localised).

Proximities and Performances

So our overall position is one in which places are seen as dynamic – 'places of movement' according to Hetherington (1997; Rojek and Urry, 1997; Ringer, 1998; Coleman and Crang, 2002b). Places are like ships, moving around and not necessarily staying in one location. They travel, slow or fast, greater or shorter distances, within networks of human and non-human agents. Places are about relationships, about the placing of peoples, materials, images and the systems of difference that they perform. We have explored how places are located in relation to material environments and objects, sand, rocks, castles, cameras, cars, fish and so on, as well as to human meanings and interactions.

But at the same time as places are dynamic, they are also crucially about proximities, about the bodily co-presence of people who happen to be in that place at that time, doing activities together, such as building a sandcastle *and* family life. There are intermittent moments of physical proximity between people that seem to be desirable or even obligatory.

Such co-presence affords access to the eyes of the other(s) (see Urry, 2003b, on the following). Eye contact enables and stabilises intimacy and trust, as well as the perception of insincerity and fear. The eye is a unique 'sociological achievement' since looking at one another is what effects the connections and interactions of individuals (Simmel, cited Frisby and Featherstone, 1997, p.111; Amin and Thrift, 2002, pp.38-9). Simmel terms this the most direct and 'purest' interaction. It is the look between people which produces moments of intimacy producing the 'most complete reciprocity' of person to person, face to face (quoted Frisby and Featherstone, 1997, p.112). The look is returned and trust can get established and reproduced. Goffman terms these 'eye-to-eye looks' (1963, p.92), while Schutz talks of 'the interlocking of glances' (1967, p.169). These co-present looks enable people to develop encounters, displaying attentiveness and commitment.

'When the eyes are joined' conversational performances flow, typically beginning with small talk (Goffman, 1963, p.92). Participants often protect the other in order not to embarrass them, and much loose talk involves helping to construct and mould the conversational performance (Boden and Molotch, 1994; Boden, 1994). We have seen how participants jointly produce conversations, even if there are significant inequalities of power and contribution involved. Face-to-face conversations are co-productions, topics come and go, misunderstandings can be corrected, commitment and sincerity are directly assessed. Trust between people is thus something that gets worked at involving a joint performance by those in such conversations, whether on the beach, walking along the harbour, in the café, or in the holiday home. Such conversations comprise words, indexical expressions, facial gestures, body language, status, voice intonation, pregnant silences, past histories, anticipated conversations and actions, turn-taking practices and so on.

Talking can also involve touch. The embodied character of conversation is thus: 'a managed physical action as well as "brain work"' (Boden and Molotch, 1994, p.262). Goffman describes how information is provided that is 'embodied' rather than 'disembodied'. Thus when 'one speaks of experiencing someone else with one's naked senses, one usually implies of the reception of embodied messages. This linkage of naked senses on one side and embodied transmission on the other provides one of the crucial communication conditions of face-to-face interaction' (Goffman, 1963, p.15).

These co-present conversations are especially important within family life, when attending family events such as weddings, christenings, marriages, funerals, Christmas, Easter and especially the shared holiday. These co-present conversations can be crucial for developing relations of trust that can persist during lengthy periods of distance and even solitude. These social obligations are associated with spending moments of 'quality time' within specific locations often involving lengthy travel away from normal sites of work and family life. There is often a quite distinct temporal feel to the moment, it is 'out of time', separate from and at odds with the 'normal' processes of work, leisure and a 'fragmented' family life ('take time out' as the advertising goes). Often as we have seen such co-presence involves obligations with material objects, to build a sandcastle on the beach, to share a meal together in the holiday home, to look through holiday photos with each other.

These encounters typically occur in specific places bodily experienced for oneself. Such 'leisure places' involve a 'face-to-place' co-presence where the body is immersed in that 'other' place, in that fishing village, by those ruins, on that beach, by that harbour. Some such places can involve 'adventure', islands of life resulting from bodily arousal, from bodies very actively in motion and often in danger (see Bell and Lyall, 2002).

There are also obligations to co-presence resulting from particular 'live' events programmed to happen at a specific moment. One event discussed above is the Roskilde Festival that attracts 70-80,000 young people from across northern Europe. Such events in real time generate intense moments of co-presence that cannot be 'missed' and produce enormous movements of people at specific moments. The bodily proximity of thousands of similar other people helps to produce a distinct atmosphere that is the very attraction of such 'alternative' 'time-compressed' events.

These co-present encounters are located within particular times and spaces, there is a 'gathering' in which people must sense that they are close enough to be seen and to see others (Goffman, 1963, p.17). People commit themselves to remain in that space for the duration of the interaction, and each uses and handles the timing of utterances and silences to perform conversations. There is an expectation of mutual attentiveness of those who are geographically and socially co-present (see Urry, 2003b).

This expectation of mutual attentiveness is by contrast with what Goffman calls the 'civil inattention' that is found amongst 'strangers' who are bodily proximate in public spaces (1963, pp.84-85; see chapter 3 above). Here we find a pattern of not attending to those others present, almost pretending that they are not present (the blasé attitude found in cities as critiqued by Simmel: Frisby and Featherstone, 1997). In holiday areas civil inattention towards others is to be found along promenades, in parks and especially on beaches where people can be more or less naked in the company of bodily proximate yet total strangers.

Mobile Methods

The significance of this book is broader than the specific studies reported. This is because we develop some distinct methods to cope with the complex patterns of presence and absence, of home and away, that holidaymaking entails. We have in other words begun to develop those 'mobile methods' to complement the new mobility paradigm that this work will hopefully promote.

Zukin has recently highlighted the complex nature of such mobile methods (2003). She analyses how her 'consumption' of a jug involves different kinds of mobile performance. First, she performs *flâneurie* and window-shopping, strolling along an area of quality European craft shops located in Manhattan. She visually consumes a particular style of jug that signifies for her good European taste, involving in part imagined travel to 'Europe'. Then later actually in Tuscany while sightseeing as a *tourist* consumer, she encounters dozens of this 'same' jug that is no longer signifies good European taste but tourist kitsch.

However, later she builds up expertise in this style of jug so developing the performance of a *connoisseur* enjoying the thrill of travel, search and acquisition, to complete the collection. Physical travel in part is organised so as to collect further examples of these particular jugs. And finally, Zukin 'travels' to eBay and becomes a commercial buyer *and* potential seller of these jugs.

Thus the apparently simple task of moving around a place, or virtually moving on the internet, and buying a jug is performed and perform-able in strikingly different ways. Zukin distinguishes between four kinds of 'mobility', flâneurie, tourist consumerism, connoisseurship and virtual travel. She shows that each of these 'produces' a different relationship with the apparently same jug. But this gets located within different circuits of anticipation, movement and memory and so is in effect a different jug consumed in a different way. Conventional social science research methods would probably be unable to grasp such distinctions between different mobile performances and hence different modes of 'consumption'.

How then to research these mobilities and to make sure that we have adequately captured the appropriate 'meanings' of such practices that are in part on the move? How to understand the role of objects in such mobile practices, something highlighted through the metaphor of the sandcastle?

One starting point from our research is that even though people travel to places of the 'familiar', there is an exceptional scale of corporeal movement even within such familiar places. People are drifting, driving, swimming, strolling, navigating, photographing. People do not stay in the one place. Hence research methods need also to be on the move, to simulate this intermittent mobility. Two methods we employed try to deal with this movement.

First, we were able to get households to keep a time diary so that they recorded for different periods what they were doing and where, and if they were moving during those periods (see chapter 7). We used this method to plot how it was the household, and indeed different household members, move through time-space. We were less interested in exactly what they did as in how they used time-space in order to construct their everyday life while 'on holiday'. These time-diaries were used to comprehend how people performed those activities often on the move, especially using the patchwork of children's drawings, mental maps and narratives that were entered in these diaries.

Second, we undertook observation and interviewing *within* particular sites, along a harbour, by the ruins of a castle, on the beach. Here we sought to participate in patterns of movement ourselves and then to interview people as to how their diverse mobilities were helping to constitute their sense of everyday life while they were away. This method of 'participation-while-interviewing' was also partly employed in the research on how photographic practices entail performances to reproduce family life. We also noted the complex ways in which photography happens. We observed (and photographed) bodies erect and kneeling, bodies bending sideways, forward and backward, bodies leaning on ruins, bodies hanging over cliffs, bodies lying on the ground, mobile bodies performing various performances of family, play and pleasure.

But not all travel is corporeal. Hugely important in mobility practices is what we have termed above 'imaginative travel'. We have seen the importance both of 'travelling' forwards in time to places only known through especially photographic images, or travelling backwards in time to places that possess haunting memories. Wordsworth referred to the latter as 'spots of time', often very short memories of a place that he says can 'retain a renovating virtue' (cited de Botton, 2002, p.154).

Imaginative travel normally involves experiencing in one's imagination the atmosphere of place. This necessitates novel research methods since atmosphere is neither reducible to the material infrastructures nor to the discourses of representation. We sought to examine the haunting atmosphere of Bornholm through lengthy interviews and with seeing the kinds of photographic images that people produced so as to capture that atmosphere of their 'day' at Hammershus. But this is never easy to research since, as Wordsworth describes in the context of the daffodils he observed along Lake Ullswater, haunting memories can occur unexpectedly. They 'flash upon that inward eye' even though one may be unaware that one is haunted by such vivid 'spots of time' (quoted de Botton, 2002, p.156). The apparently absent can dramatically and unpredictably make itself become particularly vivid.

Indeed we have seen that there is complex memory-traveling through holiday photographs that rarely contain fixed memory-stories. Memories of places visited are not simply stored for good, waiting to be consumed. Photographic images arrest the passage of time, but in people's actual *use* of them, silently or in conversation, they are enlivened and become full of time and brought to life. This necessitates the research methods we employed that simulate this active employment of photographs, although as much of this is familial or private there is a major challenge to get inside such private worlds and to discover those 'family secrets' (Kuhn, 1995).

Sometimes though these various private worlds come together in the multiple 'transfer points' that are periodically involved in being mobile. There is a large range of 'places of in-between-ness', of immobility. These include lounges, waiting rooms, cafés, amusement arcades, parks, hotels, airports, stations, motels, harbours and so on. These are 'places of encounter', places to meet, to make decisions, to wait, to socialise (and they are not non-places à la Augé, 1995; see Pascoe, 2001, Beckmann, forthcoming, on airspaces). Goffman more generally describes the 'realm of activity that is generated by face-to-face interaction and organized by norms of co-mingling – a domain containing weddings, family meals, chaired meetings, forced marches, service encounters, queues, crowds, and couples' (1971, p.13).

Such points of transfer, of co-mingling, necessitate a significant immobile network operating within that place in order that others can be on the move. To create mobility, there must be flexible networks of immobile people and materials to provide the specific mobility potential. These immobilities of those temporally stopping and those immobilised in order to move others provide opportunities for research of the sort we developed in this book. We observed tourists pausing, by the harbour, in the castle, along the beach. Thus there is the paradox that mobile methods need to be exceptionally attuned to researching the immobilities of people, places and materials.

Performing Places

In most tourism studies a clear distinction is drawn between tourists *and* those destinations that they sometimes happen to visit. The researcher typically examines those forces that from time to time propel tourists to travel to such destination places. These places are seen as pushing or pulling tourists to visit. Places are presumed to be relatively fixed, given and separate from those visiting. And the visitors in this account are taken to do qualitatively different kinds of things while they are in these destination places, compared with what they normally do 'at home'.

Our case studies in this book demonstrate many general arguments that contest this typical paradigm within tourism studies. A research programme located within what we have called the new mobility paradigm shows the inappropriateness of an ontology of qualitatively distinct 'places', on the one hand, and 'people', on the other. Our research shows that there is a complex relationality of places and peoples that are connected through diverse performances.

Thus tourist activities are not separate from the places that happen contingently to be visited. Indeed the places travelled to depend in part upon what is practised within them. Places are dynamic and depend upon performances both by 'hosts' of very many different sorts but especially by 'guests'. Moreover, many such performances are intermittently mobile 'within' the destination place itself; travel is not just a question of travelling in order to get *to* that destination. And the 'place' itself is not so much fixed but is itself implicated within complex networks by which 'hosts, guests, buildings, objects and machines' are contingently brought together so as to produce certain tourist performances in certain places at certain times. This book provides detailed qualitative accounts of various 'mobile' performances by 'tourists' that happen in certain places, performances that help to constitute those very places. These studies de-centre tourist studies away from 'tourists', but not from their patterned performances, and onto the diverse networks that contingently produce various destination places.

Places are thus (re)produced through systems of tourist performances, made possible and contingently stabilised through networked relationships with other organisations, buildings, objects and machines. Such performances can be more or less organised or formal. They include climbing, collecting, reminiscing, strolling, shopping, talking, drinking, driving, sunbathing, flâneurie, photographing, reading, dancing, eating, driving, building sandcastles and so on. Each of these we have seen involve complex practices that involve organisation of a contingent and complex network through multiple times and spaces. This is, we would suggest, the way forward for other tourism research, to leave behind the tourist as such and to focus rather upon the contingent networked performance and production of places that are to be toured and get remade as they are so toured.

Bibliography

Abercrombie, N. and Longhurst, B. (1998), *Audiences*, Sage, London.

Adler, J. (1985), 'Youth on the Road, Reflections on the History of Tramping', *Annals of Tourism Research*, Vol. 12, pp. 335-54.

Adler, J. (1989a), 'Travel as Performed Art', *American Journal of Sociology*, Vol. 94, pp. 1366-91.

Adler, J. (1989b), 'Origins of Sightseeing', *Annals of Tourism Research*, Vol. 16, pp. 7-29.

Agnew, J. (1987), *Place and Politics*, Allen & Unwin, Boston.

Ahmed, S. (1998), 'Tanning the Body: Skin, Colour and Gender', *New Formations*, no. 34, pp. 27-42.

Albers, P.C. and James, W.R. (1988), 'Travel photography: a Methodological Approach', *Annals of Tourism Research*, Vol. 15, pp. 134-58.

Amin, A. (1993), 'The Globalisation of the Economy, An Erosion of Regional Networks?', in G. Grabher (ed), *The Embedded Firm. On the Socio-economics of Industrial Networks*, Routledge, London, pp. 278-95.

Amin, A. and Thrift, N. (2002), *Cities, Reimagining the Urban*, Polity, Cambridge.

Anderson, B. (1991), *Imagined Communities*, Verso, London.

Andrews, M. (1989), *The Search for the Picturesque: Landscape, Aesthetics and Tourism in Britain, 1760-1800*, Scolar Press, Aldershot.

Andrews, M. (1999), *Landscape and Western Art*, Oxford University Press, Oxford.

Aronsson, L. (1997), 'Tourism in Time and Space: An Example from Smögen, Sweden', D.G. Lockhart and D. Drakakis-Smith (eds), *Island Tourism, Trends and Prospects*, Pinter, London and New York, pp. 118-36.

Augé, M. (1995), *Non-Places*, Verso, London.

Barba, E. (1986), *Beyond the Floating Islands*, PAJ Publications, New York.

Barthes, R. (2000), *Camera Lucida*, Vintage, London.

Batchen, G. (1999), *Burning With Desire: The Conceptions of Photography*, MIT Press, London.

Bauman, Z. (1995), *Life in Fragments. Essays in Postmodern Moralities*, Blackwell, Oxford.

Bauman, Z. (2000), *Liquid Modernity*, Polity, Cambridge.

Bauman, Z. (2001), *Community – Seeking Safety in an Insecure World*, Polity, Cambridge.

Bauman, Z. (2003), Liquid Love, Polity, Cambridge.

Beck, U. and Beck-Gernsheim, E. (1995), The Normal Chaos of Love, Polity, Cambridge.

Beckmann, J. (forthcoming), 'Ambivalent Spaces of Restlessness, Ordering (im)mobilities at Airports' in J.O. Bærenholdt and K. Simonsen (eds), *Space Odysseys - Spatiality and Social Relations in the 21st Century*.

Bell, C. and Lyall, J. (2002), 'The accelerated sublime: thrill-seeking adventure heroes in the commodified landscape', in S. Coleman and M. Crang (eds), *Tourism. Between Place and Performance*, Berghahn, New York, pp. 21-37.

Benjamin, W. (1973), *Illuminations*, Fontana, London.

Berger, J. (1972), *Ways of Seeing*, Penguin, Harmondsworth.

Boden, D. (1994), *The Business of Talk*, Polity, Cambridge.

Boden, D. and Molotch, H. (1994), 'The compulsion to proximity', in R. Friedland and D. Boden (eds), *Nowhere. Space, Time and Modernity*, University of California Press, Berkeley, pp. 257-86.

Boorstin, D. (1962), *The Image or What happened to the American Dream*, Athaneum, New York.

Booth, F. (2001), *Australian Beach Cultures, The History of Sun, Sand and Surf*, Frank Cass, London.

de Botton, A. (2002), *The Art of Travel*, Pantheon Books, New York.

Bourdieu, P. (1990), *Photography: A Middlebrow Art*, Polity, London.

Bruner, E. (1995), 'The Ethnographer/Tourist in Indonesia', in M.-F. Lanfant, J. Allcock and E. Bruner (eds), *International Tourism*, Sage, London, pp. 224-41.

Bull, M. (2001), 'Soundscapes of the Car: A Critical Ethnography of Automobile Habitation', in D. Miller (ed), *Car Cultures*, Berg, Oxford, pp. 185-202.

Butler, J.P. (1993), *Bodies that Matter, On the Discursive Limits of Sex*, Routledge, London.

Buttimer, A. (2000), 'Place Methaphor and Milieu in Hemingway's Fiction', in A.B. Murphy and D.L. Johnson (eds), *Cultural Encounters with the Environment, Enduring and Evolving Geographical Themes*, Rowman and Littlefield Publishers, Lanham, pp. 203-17.

Buzard, J. (1993), *The Beaten Track*, Clarendon Press, Oxford.

Bærenholdt, J.O. (2002), *Coping Strategies and Regional Policies – Social Capital in the Nordic Peripheries*, Nordregio Report 2002:4, Future Challenges and Institutional Preconditions for Regional Development Policy vol. 2, Stockholm.

Bærenholdt, J.O. and Aarsæther, N. (2002), 'Coping Strategies, Social Capital and Space', *European Urban and Regional Studies*, Vol. 9, pp. 151-65.

Callon, M. (1999), 'Actor-network Theory - the Market Test', in J. Law and J. Hassard (eds), *Actor Network Theory and After*, Blackwell, Oxford, pp. 181-95.

Carson, R. [1955] (2000), 'The Marginal World', in L. Lencek and G. Bosker (eds), *Beach, Stories by the Sand and the Sea*, Marlowe & Company, New York, pp. 1-8.

de Certeau, M. (1984), *The Practices of Everyday Life*, University of California Press, Berkeley.

Chalfen, R. (1987), *Snapshot Versions of Life*, Bowling Green State University Popular Press, Bowling Green, OH.

Chambers, D. (2001), *Representing the Family*, Sage, London.

Chambers, D. (2003), 'Family as Place: Family Photograph Albums and the Domestication of Public and Private Space', in J. Schwartz and J. Ryan (eds), *Picturing Place: Photography and the Geographical Imagination*, I.B. Tauris, London, pp. 96-114.

Chaney, D. (1993), *Fictions of Collective Life*, Routledge, London.

Chaney, D. (2000), 'The power of metaphors in tourism theory', in S. Coleman, and M. Crang (eds), *Tourism. Between Place and Performance,* Berghahn, New York, pp. 193-206.

Coleman, S. and Crang, M. (2002a), 'Grounded Tourists, Travelling Theory', in S. Coleman and M. Crang (eds), *Tourism Between Place and Performance*, Berghahn Books, Oxford, pp. 1-17.

Coleman, S. and Crang, M. (eds) (2002b), *Tourism. Between Place and Performance*, Berghahn Books, Oxford.

Cooper, C. (1997), 'Parameters and Indicators of the Decline of the British Seaside Resort', in G. Shaw and A.Williams (eds), *The Rise and Fall of British Coastal Resorts, Cultural and Economic Perspectives*, Mansell, London, pp. 79-191.

Corbin, A. (1994), *The Lure of the Sea, The Discovery of the Seaside in the Western World 1750-1840*, University of California Press, Berkeley and Los Angeles.

Crang, M. (1997), 'Picturing practices: Research through the Tourist Gaze', *Progress in Human Geography*, Vol. 21, pp. 359-73.

Crang, M. (1998), *Cultural Geography*, Routledge, London.

Crang, M. (1999), 'Knowing, Tourism and Practices of Vision', in D. Crouch (ed.), *Leisure/Tourism Geographies: Practices and Geographical Knowledge*, Routledge, London, pp. 238-56.

Crang, M. (2001), 'Rhythm of the City, Temporalised space and motion', in J. May and N. Thrift (eds), *Timespace, Geographies of Temporality*, Routledge, London, pp. 187-207.

Crang, M. (2002), 'Commentary: Between places: Producing Hubs, Flows and Networks' *Environment and Planning A*, Vol. 34, pp. 569-74.

Crawshaw, C. and Urry, J. (1997), 'Tourism and the Photographic Eye', in C. Rojek and J. Urry (eds), *Touring Cultures*, Routledge, London, pp. 176-95.

Cresswell, T. (ed) (2001), *Mobilities. Special Issue of 'New Formations'*, Vol. 43, Spring.

Cresswell, T. (2002), 'Introduction: theorizing place', in G. Verstraete and T. Cresswell (eds), *Mobilizing Place, Placing Mobility*, Rodopi, Amsterdam, pp. 11-32.

Crouch, D. (2002), 'Surrounded by Place, Embodied Encounters', in S. Coleman and M. Crang (eds), *Tourism between Place and Performance*, Berghahn Books, Oxford, pp. 207-18.

Crouch, D., Aronsson, L., and Wahlström, L. (2001), 'Tourist encounters', *Tourist Studies*, Vol. 1, pp. 253-70.

Crouch, D. and Lübbren, N. (2003), 'Introduction', in *Visual Culture and Tourism*, Berg, Oxford, pp. 1-20.

Debbage, K.G. (1991), 'Spatial Behaviour in a Bahamian Resort', *Annals of Tourism Research*, Vol. 18, pp. 251-68.

Debbage, K.G. and Daniels, P. (1998), 'The Tourist Industry and Economic Geograhy, Missed Opportunities', in D. Ioannides and K.G. Debbage (eds), *The Economic Geography of the Tourist Industry, A Supply-side Analysis*, London/New York: Routledge, pp. 17-30.

Degen, M. and Hetherington, K. (2001), 'Hauntings', *Space and Culture*, Vol. 11/12, pp. 1-6.

Desmond, J. (1999), *Staging Tourism: Bodies on Display from Waikiki to Sea World*, University of Chicago Press, Chicago.

Dicken, P.; Kelly, P.F.; Olds, K. and Yeung, H. W-C. (2001), 'Chains and Networks, Territories and Scales: Towards a Relational Framework for Analysing the Global Economy', *Global Networks*, Vol. 1, pp. 89-112.

Dutton, G. (1985), *Sun, Sea, Surf and Sand – the Myth of the Beach*, Oxford University Press, Melbourne.

Edensor, T. (1998), *Tourists at the Taj, Performance and Meaning at a Symbolic Site*, Routledge, London.

Edensor, T. (2000a), 'Staging Tourism: Tourists as Performers', *Annals of Tourism Research*, Vol. 27, pp. 322-44.

Edensor, T. (2000b), 'Walking in the British Countryside: Reflexivity, Embodied Practices and Ways to Escape' in P. Macnaghten and J. Urry (eds), *Body and Society*, Vol. 6, pp. 81-106.

Edensor, T. (2001), 'Performing Tourism, Staging Tourism – (Re)producing Tourist Space and Practice', *Tourist Studies*, Vol. 1, pp. 59-81.

Edwards, E. (1999), 'Photographs as Objects of Memory', in M. Kwint, C. Breward and J. Aynsley (eds), *Material Memories: Design and Evocation*, Berg, Oxford, pp. 221-36.

Ellegaard, K. (1999) 'A Time-geographical Approach to the Study of Everyday Life – a Challenge of Complexity', *GeoJournal*, Vol. 48, pp. 167-75.

Enzensberger, H.M. (1958), Vergebliche Brandung der Ferne, Eine Theorie der Tourismus, *Merkur*, Vol. 8, pp. 701-20.

Erhvervsministeriet (2000), *Regeringens Turistpolitiske Redegørelse 2000 til Folketinget*, 95-510-352, Government of Denmark, Copenhagen (Tourist Policy Report).

Erhvervsministeriet (2001), *National Strategi for Dansk Turisme – kort fortalt*, Government of Denmark, Copenhagen (Tourist Policy Report).

Eriksen, T. Hylland (2001), *Tyranny of the Moment*, Pluto, London.

Featherstone, M., Thrift, N., Urry, J. (eds) 2004), *Cultures of Automobility*, special issues of *Theory, Culture and Society* (forthcoming)

Feifer, M. (1985), *Going Places*, MacMillan, London.

Fennel, D.A. (1996), 'A Tourist Space-Time Budget in the Shetlands Islands', *Annals of Tourism Research*, Vol. 23, pp. 811-29.

Forster, E.M. [1910] (1931), *Howard's End*, Penguin, Harmondsworth.

Framke, W. (1995), 'Tourism in Denmark', in F. Vellas (ed), *An Encyclopedia of International Tourism*, Vol. I, Serdi, Paris, pp. 49-66.

Framke, W. (2001), 'Dänemark as Tourismusland: Entwicklung, Probleme und Strategien, in H. Karrasch et. al (eds), *Europa 21*, HGG-Journal Vol. 16, pp. 149-64.

Framke, W. (2002a), 'The Destination as a Concept: A Discussion of the Business-related Perspective versus the Socio-cultural Approach in Tourism Theory, *Scandinavian Journal of Hospitality and Tourism*, Vol. 2, pp. 93-108.

Framke, W. (with Sørensen, F.) (2002b), *Destination Roskilde, Netværk i produktionen af destinationer*, Publikationer fra Dansk Center for Turismeforskning, Arbejdspapir nr. 178, Publikationer fra Geografi, Institut for Geografi og Internationale Udviklingsstudier, Roskilde Universitetscenter, Roskilde.

Franklin, A. (2003), *Tourism: An Introduction*, Sage, London.

Franklin, A. and Crang, M. (2001), 'The Trouble with Tourism and Travel Theory', *Tourist Studies*, Vol. 1, pp. 5-22.

Frisby, D. and Featherstone, M. (eds) (1997), *Simmel on Culture*, Sage, London.

Gergen, K.J. (1994), *Realities and Relationships*, Harvard University Press, Cambridge Mass.

Giddens, A. (1984), *The Constitution of Society*, Polity, Cambridge.

Giddens, A. (1991), *Modernity and Self-identity*, Polity, Cambridge.

Giddens, A. (1992), *The Transformation of Intimacy*, Polity, Cambridge.

Goffman, E. (1959), *The Presentation of Self in Everyday Life*, Anchor Books, NewYork.

Goffman, E. (1963), *Behaviour in Public Places*, Free Press, New York.

Goffman, E. (1971), *Relations in Public*, Penguin, Harmondsworth.

Goffman, E. (1986), *Frame Analysis – An Essay on the Organization of Experience*, Northeastern University Press, Boston.

Green, N. (1990), *The Spectacle of Nature*, Manchester University Press, Manchester.

Gregory, D. (1994), *Geographical Imaginations*, Blackwell, London.

Gren, M. (2001), 'Time-geography Matters', in J. May and N. Thrift (eds), *Timespace, Geographies of Temporality*, Routledge, London, pp. 208-25.

Haldrup, M. (2004), 'Laid Back Mobilities: Second-home Holidays in Time and Space', *Tourism Geographies* (forthcoming).

Haldrup, M. and Larsen, J. (2003), 'The Family Gaze', *Tourist Studies*, Vol. 3, pp. 23-46.

Hall, C.M. (1994), *Tourism and Politics, Policy, Power and Place*, Wiley, Chichester.

Hall, C.M. and Page, S.J. (2002), *The Geography of Tourism and Recreation, Environment, Place and Space, Second Edition*, Routledge, London.

Harloe, M., Pickvance, C.G. and Urry, J. (eds), *Place, Policy and Politics - Do Localities Matter?* Unwin Hyman, London.

Havnegade, Et Bymiljø Projekt i Allinge, Allinge-Gudhjem Kommune, Teknisk Forvaltning (Municipal Planning Project).

Heidegger, M. (2002), 'Building Dwelling Thinking' in *Basic Writings*, Routledge, London, pp. 347-63.

Herold, E.; Garcia, R. and DeMoya, R. (2001), 'Female Tourists and Beach Boys, Romance or Sex in Tourism?', *Annals of Tourism Research*, Vol. 28, pp. 978-97.

Hetherington, K. (1997),'In place of geometry: the materiality of place', in K. Hetherington and R. Munro (eds), *Ideas of Difference*, Blackwell/Sociological Review, Oxford, pp. 183-99.

Hetherington, K. (1998), *Expressions of Identity – Space, Performance, Politics*, Sage, London.

Hirsch, M. (1997), *Family Frames: Photography, Narrative and Postmemory*, Harvard University Press, Cambridge.

Hjalager, A.-M. (2000), 'Tourism Destinations and the Concept of Industrial Districts', *Tourism and Hospitality Research*, Vol. 2, pp. 199-213.

Hjalager, A.-M. (2001), 'Den Teoretiske Referenceramme – en kort gennemgang', in A.-M. Hjalager (ed), *Nordisk Turisme i et Regionalt Perspektiv*, Nordregio Working Paper 2001:11, Stockholm, pp. 11-24.

Hoggett, P. and Bishop, J. (1986), *Organizing Around Enthusiasms*, Comedia, London.

Holland, P. (1991), 'Introduction: History, Memory and the Family Album, in J. Spence & P. Holland (eds), *Family Snaps: The Meanings of Domestic Photography*, Virago, London, pp. 2-12.

Holland, P. (2001), 'Personal Photography and Popular Photography', in L. Wells (ed), *Photography: A Critical Introduction*, Routledge, London, pp. 117-62.

Hägerstrand, T. (1983), 'In Search for the Sources of Concepts', in A. Buttimer (ed), *The Practice of Geography*, Longman, London, pp. 238-56.

Hägerstrand, T. (1985), 'Time-Geography: Focus on the Corporeality of Man, Society and Individuals', in S. Aida et al (eds), *The Science and Praxis of Complexity*, United Nations University, New York, pp. 193-216.

Håkansson, H. (1989), *Corporate Technological Behaviour. Co-operation and Networks*, Routledge, London.

Industriministeriet (1986), *Turistpolitisk Redegørelse 1986*, Government of Denmark, Copenhagen (Tourist Policy Report).

Industriministeriet (1991), *Turistpolitisk Redegørelse*, 90-420-1/CB, Government of Denmark, Copenhagen (Tourist Policy Report).

Ingold, T. (2000), *The Perception of the Environment, Essays on livelihood, dwelling and skill*, Routledge, London.

Iribas, J. M. (2000), 'Touristic Urbanism', W. Mass, J. v. Rijs and ESARQ: *MVRDV, Costa Iberica, Upbeat to the Leisure City*, ACTAR, Barcelona, pp. 106-119.

Jaakson, R. (1986), 'Second-Home Domestic Tourism', *Annals of Tourism Research*, Vol. 13, pp. 367-91.

Jackson, P. (1998), *Maps of Meaning: an introduction to cultural geography*, Unwin Hyman, London.

Jensen, B. (1993), 'Det danske Schweiz – da Bornholm blev opdaget som Turistmål', *Den Jydske Historiker*, no. 65, pp. 49-79.

Kaae, B.C. (1999), *Living with Tourism, Exploration of Destination Sharing and Strategies of Adjustment to Tourism*, PhD dissertation, The Royal Veterinary and Agricultural University and Danish Forest and Landscape Research Institute, Copenhagen.

Kaiser, A. and Andersen, T. N. (1996), *Pensionatsejere om pensionatsturisme på Bornholm*, unpublished master thesis in geography, Roskilde University, Roskilde.

Kneafsey, M. (2000), 'Tourism, Place Identities and Social Relations in the European Rural Periphery', *European Urban and Regional Studies*, Vol. 7, pp. 35-50.

Kruuse, J. (1966), *Min lykkelige barndom*, Gyldendal, Copenhagen.

Kuhn, A. (1995), *Family Secrets. Acts of Memory and Imagination*, Verso, London.

Larsen, J. (2001), 'Tourism Mobilities and the Travel Glance: Experiences of being on the Move', *Scandinavian Journal of Hospitality and Tourism*, Vol. 1, pp. 80-98.

Larsen, J. (2003), *Performing Tourist Photography*, Ph.D. thesis, Dept. of Geography and International Development Studies, Roskilde University, Roskilde.

Lash, S. and Urry, J. (1994), *Economies of Sign and Space*, Sage, London.

Latour, B. (1993), *We Have Never Been Modern*, Harvester Wheatsheaf, Hemel Hempstead.

Lefebvre, H. (1991), *The Production of Space*, Blackwell, Oxford.

Leiper, N. (1990a), 'Partial Industrialization of Tourism Systems' *Annals of Tourism Research*, Vol. 17, pp. 600-605.

Leiper, N. (1990b), *Tourism Systems, An Interdisciplinary Perspective*, Department of Management Systems, Occasional Papers 1990 no. 2, Massey University.

Leiper, N. (1993), 'Industrial Entropy in Tourism Systems' *Annals of Tourism Research*, Vol. 20, pp. 221-6.

Leiper, N. (2000), 'Are Destinations "The Heart of Tourism"'? The Advantages of an Alternative Description', *Current Issues in Tourism*, Vol. 3, pp. 364-68.

Lencek, L. and Bosker, G. (1998), *The Beach: The History of Paradise on Earth*, Pimlico, London.

Lenman, R. (2003), 'British photographers and tourism in the nineteenth century: three case studies', in D. Crouch and N. Lübbren (eds), *Visual Culture and Tourism*, Berg, Oxford, pp. 91-108.

Lennon, J. and Foley, M. (2000), *Dark Tourism*, Continuum, London.

Lenntorp, B. (1999), 'Time-Geography – at the End of its Beginning', *GeoJournal*, Vol. 48, pp. 155-8.

Littlewood, I. (2001), *Sultry Climates, Travel and Sex since the Grand Tour*, John Murray, London.

Lundvall, B.-Å. (1993), 'Explaining Interfirm Co-operation and Innovation: Limits of the Transaction-cost Approach', in G. Grabher (ed), *The Embedded Firm, On the Socioeconomics of Industrial Networks*, Routledge, London, pp. 52-64.

Lury, C. (1997), 'The Objects of Travel', in C. Rojek and J. Urry (eds), *Touring Cultures*, Routledge, London, pp. 75-95.

Lutz, C. and Collins, J. (1993), *Reading National Geographic*, University of Chicago Press, Chicago, IL.

Löfgren, O. (1999), *On Holiday, A History of Vacationing*, University of California Press, Berkeley.

Lübbren, N. (2001), *Rural artists' colonies in Europe 1970-1910*, Manchester University Press, Manchester.

Lübbren, N. (2003), 'North to South: Paradigm Shifts in European Art and Tourism', in D. Crouch and N. Lübbren (eds), *Visual Culture and Tourism*, Berg, Oxford, pp. 125-146.

MacCannell, D. (1976), *The Tourist: A New Theory of the Leisure Class*, Schocken Books, New York.

MacCannell, D. (1999), *The Tourist: A New theory of the Leisure Class* (with new epilogue), University of California Press, Berkeley.

Mackun, P. (1998), 'Tourism in the Third Italy – Labor and social-business networks', in D. Ioannides and K.G. Debbage (eds), *The Economic Geography of the Tourist Industry, A Supply-Side Analysis*, Routledge, London, pp. 256-70.

Macnaghten, P. and Urry, J. (1998), *Contested Natures*, Sage, London.

Macnaghten, P. and Urry, J. (eds) (2001), *Bodies of Nature*, Sage, London.

Maskell, P.; Eskelinen, H.; Hannibalsson, I.; Malmberg A. and Vatne, E. (1998), *Competitiveness, Localised Learning and Regional Development*, Routledge, London.

Mavor, C. (1997), 'Collecting Loss', *Cultural Studies*, Vol. 2, pp. 111-37.

McQuire, S. (1998), *Visions of Modernity: Representation, Memory, Time and Space in the Age of the Camera*, London, Sage.

Merleau-Ponty, M. (1962), *Phenomenology of Perception*, Routledge and Kegan Paul, London.

Miller, D., Jackson, P., Thrift, N., Holbrook, B. and Rowlands, M. (1998), *Shopping, Place and Identity*, Routledge, London.

Milton, K. (1993), 'Land or landscape: rural planning policy and the symbolic construction of the countryside', in M. Murray and J. Greer (eds), *Rural Development in Ireland*, Avebury, Aldershot, pp. 129-50.

Minh-Ha, Trinh T. (1994), 'Other than myself/my other self', in G. Robertson, M. Nash, L. Tickner, J. Bird, B. Curtis, T. Putnam (eds), *Travellers' Tales – Narratives of Home and Displacement*, Routledge, London, pp. 9-26.

Ministeriet for Kommunikation og Turisme (1994), *Turistpolitisk Redegørelse 1994*, Government of Denmark, Copenhagen (Tourist Policy Report).

Morgan, N. and Pritchard, A. (1999), *Tourism Promotion and Power: Creating Images, Creating Identities*, John Wiley & Sons, London.

Mowforth, M. and Munt, I. (2003), *Tourism and Sustainability: Second Edition*, Routledge, London.

Murphy, P.E. (1985), *Tourism - A community approach*, Methuen, New York.

Nilsson, P.Å. (2002), *Tourism Business Networking and restructuring on Bornholm*, Tourism Research Centre of Denmark, Working Paper no. 176, Publications from Geography, Department of Geography and International Development Studies, Roskilde University, Roskilde.

Nordtour, Roskilde Amts Turistkreds and Wonderful Copenhagen (2002), *Ferie i og omkring København*, Deskriptiv markedsanalyse og tentativ forretningsplan. (Market Analysis and Tentative Business Plan on Holiday Tourism in and around Copenhagen).

Osborne, P. (2000), *Travelling Light. Photography, travel and visual culture*, Manchester University Press, Manchester.

Ousby, I. (1990), *The Englishman's England: Taste, Travel and the Rise of Tourism*, Cambridge University Press, Cambridge.

Parinello, G.L. (2001), 'The Technological Body in Tourism, Research and Praxis', *International Sociology*, Vol. 16, pp. 205-19.

Parr, M. (1995), *Small World*, Dewi Lewis, London/Winchester.

Pascoe, D. (2001), *Airspaces*, Reaktion, London.

Paster, J. (1996), *Snapshot Magic: Ritual, Realism and Recall*, UMC, Michigan.

Pearce, D. (1988), 'Tourists Time-Budgets', *Annals of Tourism Research*, Vol. 15, pp. 106-21.

Pearce, D. (1989), *Tourism Development, Second Edition*, Longman, Essex.

Pearce, D. (1992), *Tourist Organizations*, Longman, Essex.

Perkins, H.C. and Thorns, D.C. (2001), 'Gazing or Performing', *International Sociology*, Vol. 16, pp. 185-204.

Pink, S. (2001), *Doing Visual Ethnography*, Sage, London.

Pocock, D. (1982), 'Valued Landscape in Memory: the View from Prebends Bridge', *Transactions of the Institute of British. Geographers*, Vol. 7, pp. 354-64.

Poon, A. (1993), *Tourism, Technology and Competitive Strategies*, C.A.B. International, Wallingford.

Radley, A. (1990), 'Artefacts, Memory, and a Sense of Place', in D. Middleton and D. Edwards (eds.), *Collective remembering*, Sage, London, pp. 46-59.

Rawert, O. J & Garlieb, G. (1819), *Bornholm beskreven paa en Reise i Aaret 1815*, Copenhagen.

Relph, E. (1976), *Place and Placelessness*, Pion, London.

Riles, A. (2001), *The Network Inside Out*, University of Michigan Press, Ann Arbor.

Ring, J. (2000), *How the English made the Alps*, John Murray, London.

Ringer, G. (ed) (1998), *Destinations*, Routledge, London.

Rodaway, P. (1994), *Sensuous Geographies: body, sense and place*, Routledge, London.

Rojek, C. (1993), *Ways of Escape*, Macmillan, London.

Rojek, C. and Urry, J (eds) (1997), *Touring Cultures*, Routledge, London.

Rose, G. (2003), 'Family Photographs and Domestic Spacings: a Case Study', *Transactions of the Institute of British Geographers*, Vol. 28, pp. 5-18.

Rosenquist, U. (1991), *Havnebyer og fiskerlejer, Bornholms Kystkultur*, Bornholms Turistråd, Rønne.

Rostock, X. (1927), *Badesteder i Danmark*, Turistforeningen for Danmark, Copenhagen.

Saarinen, J. (1998), 'The Social Construction of Tourist Destinations, The Process of Transformation of the Saariselkä Tourism Region in Finnish Lapland', in G. Ringer (ed), *Destinations, Cultural Landscapes of Tourism*, Routledge, London, pp. 154-73.

Saarinen, J. (2001), *The Transformation of a Tourist Destination*, Nordia Geographical Publications, Vol. 30(1), Publications of the Geographical Society of Northern Finland and the Department of Geography, University of Oulu, Oulu.

Sack, R.D. (1997), *Homo Geographicus, A Framework for Action, Awareness and Moral Concern*, Johns Hopkins University Press, Baltimore.

Said, E. (1995), *Orientalism: Western Conceptions of the Orient*, Penguin Books, Harmondsworth.

Schechner, R. (1988), *Performance Theory*, Routledge, London.

Schechner, R. (1993), 'Playing', in *The Future of Ritual – Writings on Culture and Performance*, Routledge, London.

Scheibe, K.E. (1986), 'Self-narratives and Adventure', in T. R Sarbin (ed.), *Narrative Psychology*. Praeger, New York.

Schroeder, J. (2002), *Visual Consumption*, Routledge, London.

Schutz, A. (1967), *The Phenomenology of the Social World*, Northwestern University Press, Evanston, Ill.

Schwartz, J. and Ryan, J. (eds) (2003), *Picturing Place*, I.B.Tauris, London.

Scott, A.J. (2000), *The Cultural Economy of Cities, Essays on the Geography of Image-Producing Industries*, Sage, London.

Sheller, M. (2003), *Consuming the Caribbean*, Routledge, London.

Sheller, M. and Urry, J. (2003), Mobile Transformations of 'Public' and 'Private' Life, *Theory Culture and Society*, Vol, 20, pp. 107-25.

Shields, R. (1991), *Places on the Margin, Alternative Geographies of Modernity*, Routledge, London.

Shotter, J. (1993), *Conversational Realities, Constructing Life through Language*, Sage, London.

Simonsen, K. (1996), 'What kind of space in what kind of social theory?', *Progress in Human Geography*, Vol. 20, pp. 496-512.

Simonsen, K. (forthcoming), 'Spatiality, Temporality and the Construction of the City' in J.O. Bærenholdt and K. Simonsen (eds), *Space Odysseys - Spatiality and Social Relations in the 21st Century*.

Skougaard, P. (1804), *Beskrivelse over Bornholm*. Copenhagen.

Slater, D. (1995a), 'Photography and Modern Vision: the Spectacle of Natural magic', in C. Jencks (ed.) *Visual Culture*, Routledge, London, pp. 218-37.

Slater, D. (1995b), 'Domestic Photography and Digital Culture', in M. Lister (ed), *The Photographic Image in Digital Culture*, Routledge, London, pp. 129-46.

Smart, S. and Neale, B. (1999), *Family Fragments*, Polity, Cambridge.

Smith, L. (1998), *The Politics of Focus: Women, Children, and Nineteenth-Century Photography*, St. Martin's Press, London.

Smith, M. and Duffy, R. (2003), *The Ethics of Tourism Development*, Routledge, London.

Smith, S.L.J. (1988), 'Defining Tourism, A supply-side view' *Annals of Tourism Research*, Vol. 15, pp.179-90.

Smith, S.L.J. (1991), 'The Supply-side Definition of Tourism: Reply to Leiper' *Annals of Tourism Research*, Vol. 18, pp. 312-18.

Smith, S.L.J. (1993), 'Return to the Supply-Side" *Annals of Tourism Research*, Vol. 20, pp. 226-229.

Smith, S.L.J. (1998), 'Tourism as an Industry, Debates and Concepts' in D. Ioannides and K. G. Debbage (eds), *The Economic Geography of the Tourist Industry, A supply-side analysis*, Routledge, London, pp. 31-52.

Smith, V. (ed) (1989), *Hosts and Guests. The Anthropology of Tourism*, University of Pennsylvania Press, Philadelphia.

Snepenger, D.J., Murphy, L., O'Connel, R. and Gregg, E. (2003), 'Tourists and residents use of a shopping space', *Annals of Tourism Research*, Vol. 30, pp. 567-80.

Solnit, R. (2001), *Wanderlust. A History of Walking*. Verso, London.

Sontag, S. (1977), *On Photography*, Penguin Books, London.

Spence, J. & Holland, P. (1991) (eds.), *Family Snaps: The Meanings of Domestic Photography*, Virago, London.

Squire, S.J. (1994), 'Accounting for Cultural Meanings: the Interface between Geograpghy and Tourism Studies Re-examined', *Progress in Human Geography*, Vol. 18, pp. 1-16.

Storper, M. (1995), 'The Resurgence of Regional Economies, Ten Years Later: the Region as a Nexus of Untraded Interdependencies', *European Urban and Regional Studies*, Vol. 2, pp. 191-221.

Suadicani, H. (2002), *Dansk Kystzone-landskab og Forvaltning*, PhD dissertation, Department of Environment, Technology and Social Studies, Roskilde University.

Suvantola, J. (2002), *Tourist's Experience of Place*, Ashgate, Aldershot.

Szerszynski, B. and Urry, J. (2002), 'Cultures of cosmopolitanism', *Sociological Review*, Vol. 50, pp. 461-81.

Sørensen, F. (2002a), *Destination Jammerbugten* - 'Networks in the production of destinations', Dansk Center for Turismeforskning, Arbejdspapir nr. 175, Publikationer fra Geografi, Institut for Geografi og Internationale Udviklingsstudier, Roskilde Universitetscenter, Roskilde.

Sørensen, F. (2002b), 'Tourist Destination Networks - Networks of Agglomerations as Sub-optimal Network Configurations', paper presented at the Nordic Research School on Local Dynamics PhD course, Iceland, 9-15 September.

Thrift, N. (1996), *Spatial Formations*, Sage, London.

Thrift, N. (2001), 'Still life in nearly present time', in P. Macnaghten and J. Urry (eds), *Bodies of Nature*, Sage, London, 34-57.

Tinsley, R. and Lynch, P. (2001), 'Small tourism business networks and destination development', *International Journal of Hospitality Management*, Vol. 20, pp. 367-78.

Tress, G. (1999), *Die Ferienhauslandschaft, Motivationen, Umweltsauswirkungen und Leitbilder in Ferienhaustourismus in Dänemark*, PhD dissertation, Research Report no. 120, Department of Geography and International Development Studies, Roskilde.

Tress, G. (2002), 'Development of Second-Home Tourism in Denmark', *Scandinavian Journal of Hospitality and Tourism*, Vol. 2, pp. 109-22.

Tuan, Y.F. (1976), *Topophilia: a Study of Environmental Perception, Attitudes and Values*, Prentice-Hall, Englewood Cliffs.

Tuan, Y.F. (1996), 'Space and Place: Humanistic Perspective', in. J. Agnew, D.N. Livingstone and A. Rogers (eds), *Human Geography – an essential anthology*, Blackwell, Oxford (originally published 1974 in *Progress in Geography*, Vol. 6, pp. 233-46).

Turner, V. (1985), *Dramas, Fields and Metaphors: Symbolic Action in Human Society*, Cornell Univeristy Press, Ithaca.

Urbain, D. (1994), *Sur la plage: Moers et coustumes balnéaires*, Payot, Paris.

Urry, J. (1990), *The Tourist Gaze*, Sage, London.

Urry, J. (1995), *Consuming Places*, Routledge, London.

Urry, J. (2000), *Sociology beyond Society, Mobilities for the 21st century*, Routledge, London.

Urry, J. (2002), *The Tourist Gaze, Second Edition*, Sage, London.

Urry, J. (2003a), *Global Complexity*, Polity, Cambridge.

Urry, J. (2003b), 'Social networks, travel and talk', *British Journal of Sociology*, Vol. 54, pp. 155-76.

Veijola, S. & Jokinnen, E. (1994), 'The body in tourist studies', *Theory, Culture and Society*, Vol. 6, pp. 125-51.

Verstraete, G. and Cresswell, T. (eds) (2002), *Mobilizing Place, Placing Mobility*, Rodopi, Amsterdam.

Walton, J.K. (1983), *The English Seaside Resort, A Social History 1750-1914*, St Martin's Press, New York.

Wang, N. (1999), 'Rethinking Authenticity in Tourism Experience', *Annals of Tourism Research*, Vol. 26, pp. 349-70.

Wearing, B. and Wearing, S. (1996), 'Refocusing the Tourist Experience: the "Flâneur" and the "Choraster"', *Leisure Studies*, Vol. 15, pp. 229-43.

Wells, L. (2001), *Photography: A Critical Introduction*, Routledge, London.

West, N. (2000), *Kodak and the Lens of Nostalgia*, University of Virginia Press, Charlottesville.

Wilde, O. [1891] (1951), *The Picture of Dorian Gray*, Penguin, London.

Williams, A. M. and Shaw, G. (1998), 'Tourism Policies in a Changing Economic Environment', in A.M. Williams and G. Shaw (eds), *Tourism and Economic Development, European Experiences*, Wiley, Chichester, pp. 375-91.

Williams, D.R. and Kaltenborn, B.P. (1999), 'Leisure Places and Modernity – The Use and Meaning of Recreational Cottages in Norway and the USA', in D. Crouch (ed), *Leisure/Tourism Geographies – practices and geographical knowledge*, Routledge, London, pp. 214-30.

Williams, R. (1972), 'Ideas of nature', in J. Benthall (ed), *Ecology. The Shaping Enquiry*, Longman, London, pp. 146-66.

Zukin, S. (2003), 'Home shopping in the global marketplace', Paper to *Les sens de mouvement colloque*, Cerisy-la-salle, June.

Økonomi- og Erhvervsministeriet (2002), *Dansk Turisme – Handlingsplan for Vækst*, Government of Denmark, Copenhagen (Tourist Policy Report).

Index